Napoleon, the Ice Age, and the Empire Sun
Ice Age Science Illustrated, by Rolf A. F. Witzsche

© Text Copyright Rolf A. F. Witzsche 2017
all rights reserved

Napoleon's retreat by Vasily Vereshchagin - photo by Shakko - Own work, CC BY-SA 3.0, https://commons.wikimedia.org/w/index.php?curid=25006628

The book contains the transcript with images of the Ice Age science exploration video with the above title by Rolf A. F. Witzsche.

The exploration is presented in the following 3 parts that flow into each other:

Part 1: Napoleon's catastrophic defeat in Russia.

Napoleon's defeat in Russia can be seen as a parallel event (in the small) of the Ice Age Challenge before us, which promises to defeat humanity more extensively than Napoleon was defeated, if it fails to recognize the nature and the magnitude of the Ice Age that already looms on the near horizon, potentially as near as the 2050s. Napoleon invaded an 'unknown' country without the infrastructures prepared to support a large army in a primitive landscape. Of the 640,000 troops that followed Napoleon and the empire flag into Russia with dreams of glorious victories and looting, never imagined that they were on a death march that almost no one would survive. Humanity is on the same type of track on its march into the next Ice Age. No one has any idea of how to live without agriculture that our food supply depends on, while at the same time no new infrastructures are envisioned, much less planned or being build, with which to continue agriculture in a collapsing climate towards the phase shift that turns much of the Earth as we know it, into an Ice Planet Earth.

The collapse of the system of empire, that empire forced upon itself in 1812, began with Napoleon's catastrophic defeat in Russia. Napoleon's rush into Russia with an empty dream, prefigures humanity of today rushing equally blindly into its near Ice Age future, with no preparations for it.

- Yes, the recurring Ice Age is our future (in the 2050s). This cannot be avoided. While the reality can be ignored, as it is by a society with closed eyes and a closed mind, the ensuing effects cannot be avoided. Only the consequences of the effects can be avoided by society building itself a higher-level world that the Ice Age climate cannot affect.

Part 2: The Empire Sun: Its false face.

Empire bears a false face. It claims the glory of being a sun, while the opposite is its nature. In so doing, empire also projects a false image onto the Sun itself that is totally contrary to the Sun's very nature. Empire models the Sun after itself. But is the universe defined by empire, or is it defined by its own principles?

- Physical evidence colors the Sun with the colors of the truth. By humanity knowing the truth, it can avoid the tragic consequences of chasing after dreams. When we move with demonstrable scientific proof, we move with the dynamics of the universe, not against them.

Part 3: The dawn of freedom, from empire.

1813 saw the dawning of the 'end of empire' in Europe by the dawn of a new breath of freedom. With freedom of the truth in the heart, humanity writes itself a ticket for economic development to happen, with which to meet the common needs and aims of mankind.

- The Ice Age Challenge, when taken up, reveals immense development potentials. On this path the system of empire falls by the wayside as humanity reaches for the 'stars' in building itself a New World with new industrial revolutions, and advanced platforms for human living.

Contents

Part 1: Napoleon's catastrophic defeat in Russia 13

    The Cygni Ice Age Climate 'University' 15

        ➢ *Start of Exploration* 16

    Napoleon's Grande Armée invaded to force Russia's surrender 16

    The British ruled the seas. With it, they owned India. 17

    The official excuse was to liberate Poland 18

    Militarily, the invasion was deemed to be a cake walk. But the Russians had refused to fight. 19

    Blinded by his own ideology, the ideology of empire, Napoleon walked into a trap 20

    The long train of heavy wagons forged deep spoors in the mud 21

    Some say that it was mostly the Russian winter 22

    By the time Napoleon had taken Moscow, he had fewer than 100,000 men remaining 23

    The cold may have killed half of the soldiers that eventually vacated Moscow 24

    Some say that 1,000 to 4,000 soldiers had made it back home 25

    Tchaikovsky's great 1812 Overture to celebrate Russia's victory 26

    Napoleon was not defeated by heroics, but by a dead poet and the quiet rustle of lice 27

    Nevertheless, the Russian victory was not a passive victory 28

    Napoleon was defeated by the ideology of empire itself 29

        ➢ *Empire is a mistake* 30

    The people who strode with Napoleon all ended up dead 31

    You Are on that bridge with them 32

    Humanity follows the flag of Empire, oblivious of the ice age before it 33

    Empire parades itself to be a Sun, a Star of Glory that shines by its own power ... 34

    Empire: a system of false values 35

    Empire is a force directed against humanity by which empire dooms itself 36

    Empire is also the enemy of science 37

    Our astrophysical Sun, is being paraded on posters with a false face 38

But there is light at the end of the tunnel symbolized by the cannon fire 39

The Empire Sun poster is strung across the landscape of science 40

Empire has devised a model for the Sun that mirrors its own image 41

Empire paints the world with the colors of its ideology 42

Every great artist paints the images of the world as they appear in the artist's mind 43

The masters of empire who control the sciences are steeped in the ideology they serve 44

Living oblivious of the truth the Ice Age is being swept under the rug 45

Oblivious humanity is poised to re-experience the doom of this folly 46

That this fate is near, is evident by the fact that the infrastructures are not being built 47

- Light at the end of the tunnel 48

The Real Sun is different from the Empire Sun in every respect 49

On its surface, the largely empty Sun is teeming with energetic electric interactions. 50

Wherever electric movements happen, magnetic phenomena follow 51

One of the Sun's smallest features closes the door on the Empire Sun model 52

Whether the sunspots are small in size or large, all share one common characteristic 53

No matter how closely we look, and how much we expand the view 54

Throughout the universe, 99.999% of all mass exists in the form of plasma 55

Quarks are understood as but moving points of energy 56

Explicate features of the implicate order of the energy background of cosmic space 57

That's how the great theoretical physicist David Bohm had perceived the universe 58

The specific arrangements of the quarks are recognized 59

With the same force, particles with equal charge repel one another 60

At very close distances, a strong nuclear force repels the two apart 61

On this basis plasma becomes bound up into atoms 62

When plasma exists in unbound form in dense concentration 63

Brown dwarf stars are typically of the size of Jupiter 64

Extremely large concentrations of plasma, gravity begins to play a role too 65

- ➢ A sun must be 'hollow' to function 66

The migration of the electrons away from the center of gravity 67

A sphere of atomic gas, in contrast, lacks the electric features of plasma 68

A sphere of atomic elements has its greatest mass density at its core 69

The resulting density gradient renders a gas sphere unsuitable for being a sun 70

In comparing Jupiter with Saturn, Jupiter, which has double the volume of Saturn 71

The Sun cannot be a gas sphere, but is an empty shell of plasma. with nothing substantial inside of it? 72

If we consider the extremely large star UY Scuty, the largest star 73

Not possible under the gas-compression nuclear-fusion-sun model 74

- ➢ The plasma 'sunshine' onto the Sun 75

Just as streams of sunlight flood the earth, and illumine the landscape 76

UY Scuti is immensely luminous as a huge canvas that has plasma 'shining' onto it 77

We need a university type of approach to be able to see with the mind 78

I chose the name Cygni for my own 'university' type explorations 79

The Empire-Sun poster has the opposite effect 80

Part 2: The Empire Sun: Its false face 81

- ➢ The Empire Sun: a trap 82

The effect of this trap is, that it reduces the cognitive power of humanity. 83

Mysticism that all structures of empire are built on, such as the later illuminaticism 84

Empire creates conflicts in the mind for which no solutions are deemed possible 85

On the political front, wars are achieved on the reductionist basis 86

- ➢ Conspiracy for a false Sun, hiding reality, obscures that the Ice Age transition has begun 87

The Empire Sun model is of the latter type type, that of the conspiracies 88

On a much higher-level platform than the reductionist platform that Empire imposes 89

The plasma pinching continues until the magnetic fields tangle up, flip backwards 90

The concentrated plasma, under the confinement dome, becomes focused onto a sun 91

We can see the complementary nature of the primer fields clearly evident 92

That the Primer Field's principle applies to our Sun is evident in the measurements ... 93

The voids were encountered exactly where the experimental models indicate ... 94

The void that is encountered in space by Ulysses, accords with the result ... 95

The effect that causes the Sun to be located at a node point ... 96

In plasma, electrons and protons are free flowing ... 97

Atoms are electrically balanced structures of plasma that become electrically neutral ... 98

The synthesis of atoms creates the vital sink effects ... 99

➢ Sunlight NOT from the Sun ... 100

The atoms that are forged on the surface of the Sun ... 101

It takes a wide-ranging assortment of different atomic elements ... 102

This 'perfect' balance isn't possible under the Empire Sun model ... 103

Of course, the Plasma Sun emits not only atoms and light ... 104

The solar wind is comparable to a kettle letting off steam ... 105

The surface of the Sun is a vast sea of 'granular' fusion cells ... 106

The excess pressure is released in concentrated form ... 107

The expanding plasma becomes the solar wind that flows away from the Sun ... 108

The solar wind is able to flow out through the solar corona ... 109

The solar fusion cells also emit cosmic rays ... 110

While most of the escaping cosmic rays get trapped in the solar corona ... 111

The science researcher Simon Snoll had conducted a series of reaction-experiments ... 112

The measured results always varied ... 113

Cosmic rays also affect our atmosphere, in almost the same manner ... 114

Another indicator that the Sun is getting weaker ... 115

➢ What masters the Sun? ... 116

Numerous forms of evidence prove that the Sun is not its own master ... 117

All of these variable factors, which affects our Sun from areas deep in space ... 118

Our Galaxy is presently at its weakest state in 440 million years ... 119

The glaciation conditions get interrupted periodically with warm interglacial periods 120

The solar system is not as simplistic as I had illustrated earlier 121

When the interstellar plasma stream is strong, the flow volume is large enough 122

When one set of the primer fields no longer forms 123

At the low-power level the surface temperature of the Sun drops down 124

For us on Earth, the difference will be felt like a shock 125

➢ The Sun is in a free fall 126

The timing of the Ice Age start up becomes critical for humanity 127

The ice core records tell us that our world has been getting progressively colder 128

The plasma fusion reactions on the surface of the Sun 129

The escaping cosmic rays impact the Earth atmosphere and cause 130

The measurements of the ratio in historic samples 131

When the interstellar plasma is dense, the plasma sphere around the Sun is dense likewise 132

The resulting trends match the known historic climate trends 133

The climate spikes of the 3 warming periods that occurred during the last 3,500 years 134

No matter what had caused the up-ramping of the Sun in the 1700s 135

➢ Science lets the evidence stand to speak for itself 136

The rate of the collapse has been measured by the Ulysses spacecraft 137

With the solar cosmic-ray flux sharply increasing, under the weakening Sun 138

The rate of solar collapse that the Ulysses spacecraft saw still continues 139

We have laboratory-developed proof that when a fusion cell is clogged up 140

The laboratory experiment also proves that the Empire Sun is an impossible dream 141

We may see the building of the floating infrastructures to begin soon 142

In order to make the full relocation of humanity into the tropics possible 143

One of the chief blocking factors that prevents the Ice Age recognition 144

Under the invariable Sun all climate changes are said to be necessarily man-made 145

The orbital cycles theory, termed the Milankovitch Cycles theory 146

The Milankovitch Cycles theory has been attached to the Empire Sun doctrine 147

If the Empire Sun is retained 148

That's the significance of the Empire Sun poster 149

The resulting radical depopulation that the false poster sets the stage for 150

Part 3: Dawn of freedom, from empire 151

> Love versus Empire - the beginning of the end of Empire 152

While Napoleon wasn't aiming directly for genocide with his campaign 153

The genocidal objective that started with the doctrine of the Empire Sun, has become a monster 154

Napoleon was a saint in comparison with modern society 155

Napoleon didn't intend the tragedy that he caused to happen 156

In contrast with Napoleon, who wanted to develop a land bridge to India 157

Napoleon's troops met their ice age that almost none survived 158

We, in our time, are in need to build a World Bridge with Africa at its center 159

Napoleon's soldiers would cry out to us from their icy grave if they could 160

> Divine Love, reflected in universal love for one another 161

In 1812 the soldiers did not wield their swords out of love for humanity, but to kill 162

It may have been the frivolous waste of the 680,000 of Europe's most able men 163

Miraculously, Napoleon managed to mobilize a new army against the Alliance 164

The French-Empire forces were already loosing before the big battle happened 165

The Battle of Leipzig, in Germany, started a week later 166

The battle was brought into the city 167

The position at Leipzig afforded major advantages to Napoleon's army 168

Several rivers converged and split the surrounding terrain into separate sectors 169

The Allies had difficulty moving large numbers of troops into Napoleon's sectors 170

But the momentum was no longer on Napoleon's side 171

After 4 days of war, on October 19, Napoleon ordered the retreat 172

History records little of the details, of the struggles, hopes, sacrifices 173

History also records that the vast French Empire existed no more 174

The Empire lost this day all of its territories east of the Rhine 175

The liberated people who had won their freedom from Empire 176

Construction began 84 years later in 1898. The work was completed in 15 years. 177

The great monument of freedom from Empire is constructed in granite-faced concrete 178

The power that dissolves Empire is the humanity of the human being. 179

The monument is also known for the wide view that it offers to visitors 180

The year 1812 marks the beginning of the end of the system of Empire 181

One day, however, and hopefully soon, the deadly poster of the Empire Sun 182

Till then, the poster of the Empire Sun still rules and humanity remains in crisis 183

However, times are changing 184

If the G20 logo symbolizes solar plasma-flow physics, it would indicate 185

> Ice Age = Opportunities 186

Whether humanity succeeds on this front remains yet to be seen 187

If humanity fails to break itself away from the still widely accepted doom 188

The time has come for humanity to take the trash out of the house 189

Its humanity still stands 190

Its aspiration for freedom still stands 191

Its vision still stands and continues to enrich the landscape 192

The victory that it commemorates was won in 4 decisive days in 1813 193

The Statue of Liberty in New York harbour stands on the same eternal foundation 194

The copper statue was donated as a gift from the people of France 195

> Where the heart smiles, there is liberty 196

The pedestal on which the statue stands was paid for by 120,000 individual contributions 197

This beginning of the end of Empire may have been recognized already in 1813 198

The Statue of Liberty prefigures the inevitable recognition of the inherent dignity 199

> We are not impotent - this is true now 200

While the word, 'Ice Age,' is no longer spoken, as it falls outside the framework	201
The Ice Age Challenge before us should be the banner headline in today's world	202
The masters of empire, in contrast, still call it a crime	203
The critical success in this arena depends on whether humanity raises itself up	204
➢ Arise and Shine our time has come - humanity=infinity	205
Like seeds are we, wind-blown	206
Harvest is seedtime, thoughts ripening	207
Thoughts do awaken	208
Who owns the seeds? Do we?	209
Thoughts are seeds, becoming ideas	210
Like seeds, thoughts fall to the ground	211
Love for one-another, the human spring	212
Thoughts are the Universe unfolding	213
Each harvest is seedtime	214
Builders of worlds are we	215
Who owns the cradle for the seed?	216
The melody of nature - what a song!	217
Isten to the song	218
Listen to the symphony of our humanity	219
➢ More from the author:	220
14 Libraries of books and video productions	220

Part 1: Napoleon's catastrophic defeat in Russia

➢ Napoleon, Ice Age, and the Empire Sun

Napoleon, the Ice Age, and the Empire Sun, is a Rolf A. F. Witzsche Exploration Production presented in 3 parts flowing into 1, published by Cygni Communications Ltd. BC, Canada

Part 1:
It deals with the collapse of the system of empire in Europe in 1812-1813 following Napoleon's catastrophic defeat in Russia. Will we be defeated by the Ice Age in our time, in the 2050s or sooner, for similar reasons?

Part 2
The false face of empire stands behind the false image of the Sun. Physical evidence paints with the colors of truth. By knowing the truth, we can avoid the Ice Age consequences, and move with the dynamics of the universe, not against them.

Part 3:
The 1813 beginning of the 'end of empire' in Europe became the dawn of freedom. With this freedom in the heart a new age of humanity and development was dawning. This can happen again. The Ice Age Challenge opens immense potentials. Living in a world without empire we reach for the 'stars' in building new worlds with industrial revolutions, and creative cultural living.

The Cygni Ice Age Climate 'University'

From the Cygni Ice Age Climate 'University'

Exploring an alternate theory for the Ice Ages, and of the Sun as the cause for them.

I am presenting a new video production, in honour of the "Belt and Road" world forum in Beijing, China, on May 14-15, 2017, designed as a win-win platform for economic development - peace and security building on the foundation of international cooperation and universal prosperity. My video explores the foundation on which the project may ultimately succeed.

The 'cake walk' by Napoleon into Russia, in 1812, became a disaster that killed the Grand Army and subsequently sank the French Empire. 680,000 soldiers walked to their death, oblivious that the glory of empire is not a sun, but a trap. Humanity is on the same track, oblivious of the near Ice Age, trapped by a false theory of the Sun that was created by empire in its image. Will still have a chance to follow the 1813 path (implemented in Europe) to freedom.

The music for this video has been selected in honouring the historic significance of the Palmyra concert. See: http://www.ice-age-ahead-iaa.ca/ice_age/Hollow_Sun_Napoleon.html

Rolf Witzsche

> *Start of Exploration*

Napoleon's Grande Armée invaded to force Russia's surrender

On the 24th of June 1812, Napoleon's Grande Armée of 420,000 men, crossed the Niemen River into Russia. It was to be a cake walk. It became a disaster instead.

They came to defeat the Russian army; to force Russia's surrender. wanted a land-route to conquer India.

The British ruled the seas. With it, they owned India.

The British ruled the seas. With it, they owned India. Napoleon wanted India for the French. For this, he wanted to compel Tsar Alexander, who was legally an ally of France, to give up his control of Russia to him.

 After Napoleon had defeated the Russian army in 1807 in the Battle of Friedland, France and Russia had signed the Treaties of Tilsit that had made the two countries allies. The treaty prohibited Russia from doing business with Britain, which Russia ignored. The breach enabled the objective for regime change to be forced.

The official excuse was to liberate Poland

The Grande Armée crossing the Niemen River into Russia, June 1812

The official excuse for the invasion, of course, in typical empire fashion, was that the invasion of Russia was necessary to liberate Poland from the Russian threat.

Militarily, the invasion was deemed to be a cake walk. But the Russians had refused to fight.

Napoleon and his staff at Borodino by Vasily Vereshchagin

Militarily, the invasion was a cake walk. Russia had no military might to speak of, with which to counter the 'civilized' forces of the French Empire. It was a cake walk also, because the Russians had refused to fight. They would have been slaughtered. Instead they retreated. Their refusal became their liberation.

The war was decided by logistics, not heroics. Napoleon lost the war by the lack of logistics.

The massive invasion encountered a land of poverty with squalid living conditions and teeming with lice. Lice are carriers of the typhoid. Napoleon lacked the logistics to counter this force.

During the first month into the invasion, the loss of men from disease and desertion had reduced Napoleon's army to half, so that his advisors urged Napoleon to withdraw. He agreed, but two days later changed his mind. With his eyes set on Moscow, he proclaimed. "Victory will justify and save us."

An alert person would have seen the writing on the wall.

Napoleon was defeated by the system of empire, a system of 'civility' that disregards the human dimension, and the common aim of all mankind. For empire, the goal is stealing. Wars are fought to open the doors to stealing. The Russian gave him a burned landscape that contained no food, lacked water, and offered no fodder for the horses either.

Blinded by his own ideology, the ideology of empire, Napoleon walked into a trap

Russian Cossacks against Napoleon - F. de Myrbach — wikipedia

Blinded by his own ideology, the ideology of empire, Napoleon walked into a trap that he could not see, nor his advisors could see, who were similarly blinded.

The Russians won the war by simply drawing Napoleon deep into the country while disrupting the supply trails. The Russian defenders literally watched the Grand Army to defeat itself with diseases and hunger. They literally watched it starve itself to death. The transportation infrastructure didn't exist to support the supply system for a large army. Roads, frequently became rivers of mud.

The long train of heavy wagons forged deep spoors in the mud

Trail of Attrition - a testament of the nature of empire -

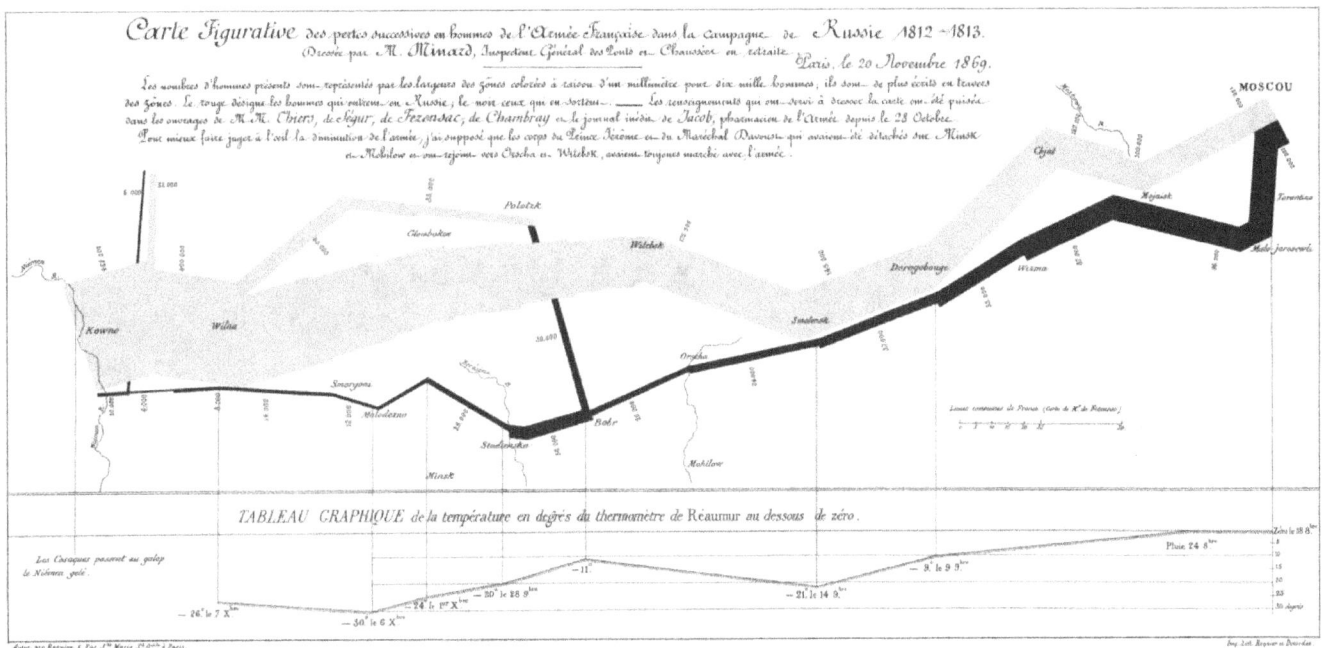

The long train of heavy wagons forged deep spoors in the mud. Wagons sank to their axels into the road. And on days when the Sun broke through, the mud became hard. The spoors became canyons. The horses broke their legs in them.

At the end of the 1st week, the Grand Army had left behind along the road a trail of 10,000 dead horses.

Some say that it was mostly the Russian winter

Some say that it was mostly the Russian winter with its minus-30 to 40-degrees temperatures and high winds, that had defeated Napoleon. This is not true.

By the time Napoleon had taken Moscow, he had fewer than 100,000 men remaining

The burning of Moscow - 1812

By the time Napoleon had taken Moscow, he had fewer than 100,000 men left out of the more than 600,000, that he had brought into Russia. And for what? His victory had been denied. He had conquered an empty and undefended city that was set ablaze within days under his very nose to deny him the facilities of civilization.

Most likely it wasn't the burning of Moscow that eventually forced his retreat. The collapsing logistics and the rising pandemics may have impelled that. Napoleon had received 15,000 reinforcements to bolster his position in Moscow. Of these 10,000 died of diseases, within the destroyed city.

The cold may have killed half of the soldiers that eventually vacated Moscow

Napoleon's retreat in Russia, 1812 - Adolph Northen.

The cold may have killed half of the soldiers that eventually vacated Moscow. Without protective logistics it doesn't matter whether the cold drops to minus 20 or 40 degrees. The Grand Army had come in summer clothing. The worst fate was for a man to fall asleep in the icy winds and in 30-below weather with driving snow. Those who surrendered to sleep, never woke up. No one had the strength left by then to help others along. The road home became littered with their frozen corpses.

Some say that 1,000 to 4,000 soldiers had made it back home

Some say that 1,000 to 4,000 soldiers had made it back home as able men, of the 100,000 who retreated from Moscow. The rest fell by the wayside.

This means that almost no one had survived the six-months long epic of dying , of the 680,000 men who had waged war on Russia.

Tchaikovsky's great 1812 Overture to celebrate Russia's victory

Tchaikovsky's great 1812 Overture, commissioned to celebrate Russia's victory, ends with the boom of cannon fire incorporated into the heroic-style music. But this isn't how the war was won.

Napoleon was not defeated by heroics, but by a dead poet and the quiet rustle of lice

Trail of Attrition - a testament of the nature of empire -

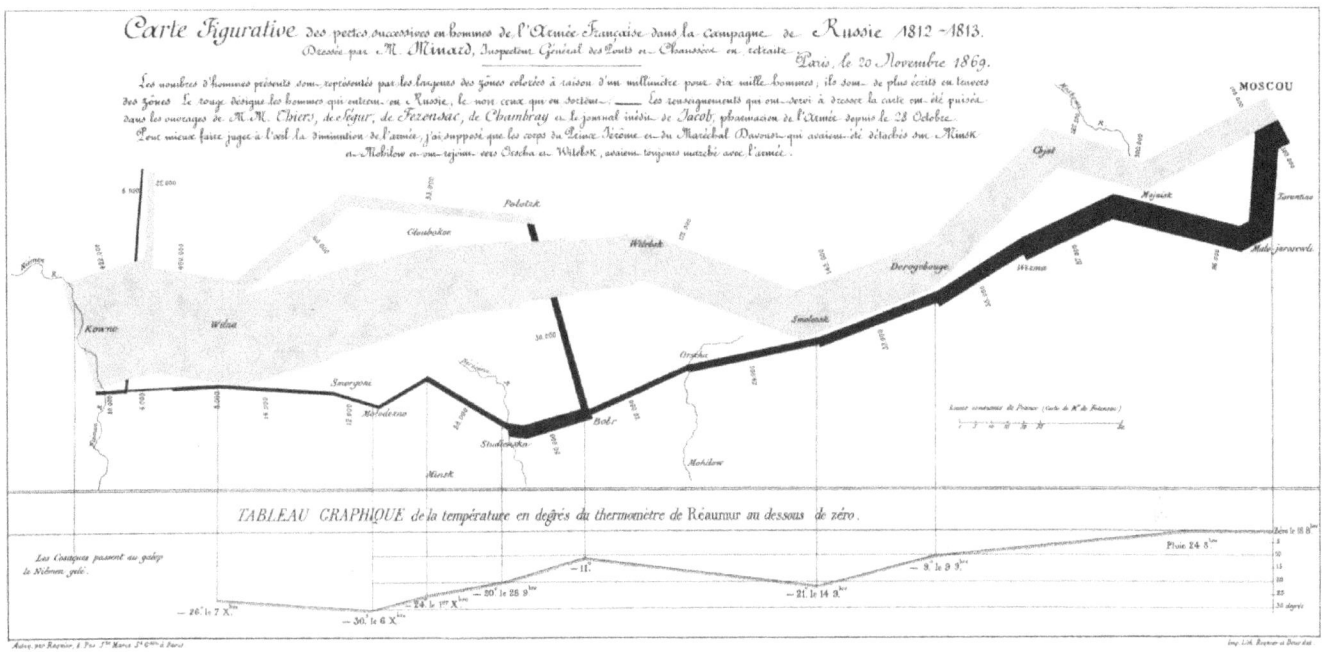

Napoleon was not defeated by heroics, but by a dead poet and the quiet rustle of lice, the growls of empty stomachs, and the freezing of soldiers who fell asleep standing.

Nevertheless, the Russian victory was not a passive victory

Nevertheless, the Russian victory was not a passive victory. It was an active victory right from the start that was designed to protect the most precious asset that a nation has, which is its people. The capital city was sacrificed to protect the most precious. 'A city can be rebuild when the nation is preserved,' was the heart of the message of the poet. The cannon-fire of the celebration symphony likely do not represent the explosions of war, but the exploding fire of humanist ideals, ideals of truth, that end wars.

Napoleon was defeated by the ideology of empire itself

The Coronation of Napoleon in 1804 by Jacques-Louis David

Mostly, however, Napoleon was defeated by the ideology of empire itself that had inspired the invasion of Russia in the first place. Napoleon was defeated by his blindness, that was symbolized by the crown that he wore. He could not see the human dimension that is blocked by this barrier of gold. Blind as he was, he saw a military victory within reach. Had he read the writing of Friedrich Schiller, the German poet of freedom, he would have chosen a different type of victory, the victory of economic development in cooperation with Russia, instead of the victory of the grave that buried his finest men and his opportunity to make his name a great and honourable one.

Napoleon had denied himself and the people of Western Europe the only possible type of real victory that is possible, built on advancing the most precious that a society has, which is the creative and productive capacity of the human being. The Russian leadership chose this type of victory - a victory that is inherently possible by the principle of our humanity.

Ultimately, victory is not possible on the side of empire, which is itself but a fading dream.

➢ **Empire is a mistake**

Empire is a mistake

- not a nation, or person, or institution -
but a false ideology, a dream, a void

Empire is a mistake
It is not a nation, or person, or institution -
but a false ideology, a dream, a void.

The people who strode with Napoleon all ended up dead

The people who strode with Napoleon all ended up dead, defeated by the myth of Empire as the rising sun.

You Are on that bridge with them

The Grande Armée crossing the Niemen River into Russia, June 1812

**You Are on that bridge with them
The whole of humanity is on that bridge**

** You Are on that bridge with them

The whole of humanity is on that bridge.

Humanity follows the flag of Empire, oblivious of the ice age before it

Humanity follows the flag of Empire, oblivious of the ice age before it, which will be its doom, because the infrastructures are presently not being built that support human living in an ice age world.

Empire parades itself to be a Sun, a Star of Glory that shines by its own power

The Coronation of Napoleon in 1804 by Jacques-Louis David

Empire is deadly. It is a belief in something that is not real. Empire parades itself to be a Sun, a Star of Glory that shines by its own power and merit. It pretends to symbolize grandeur, riches and civility, while unseen, at its core, it is devoid of humanity.

Empire: a system of false values

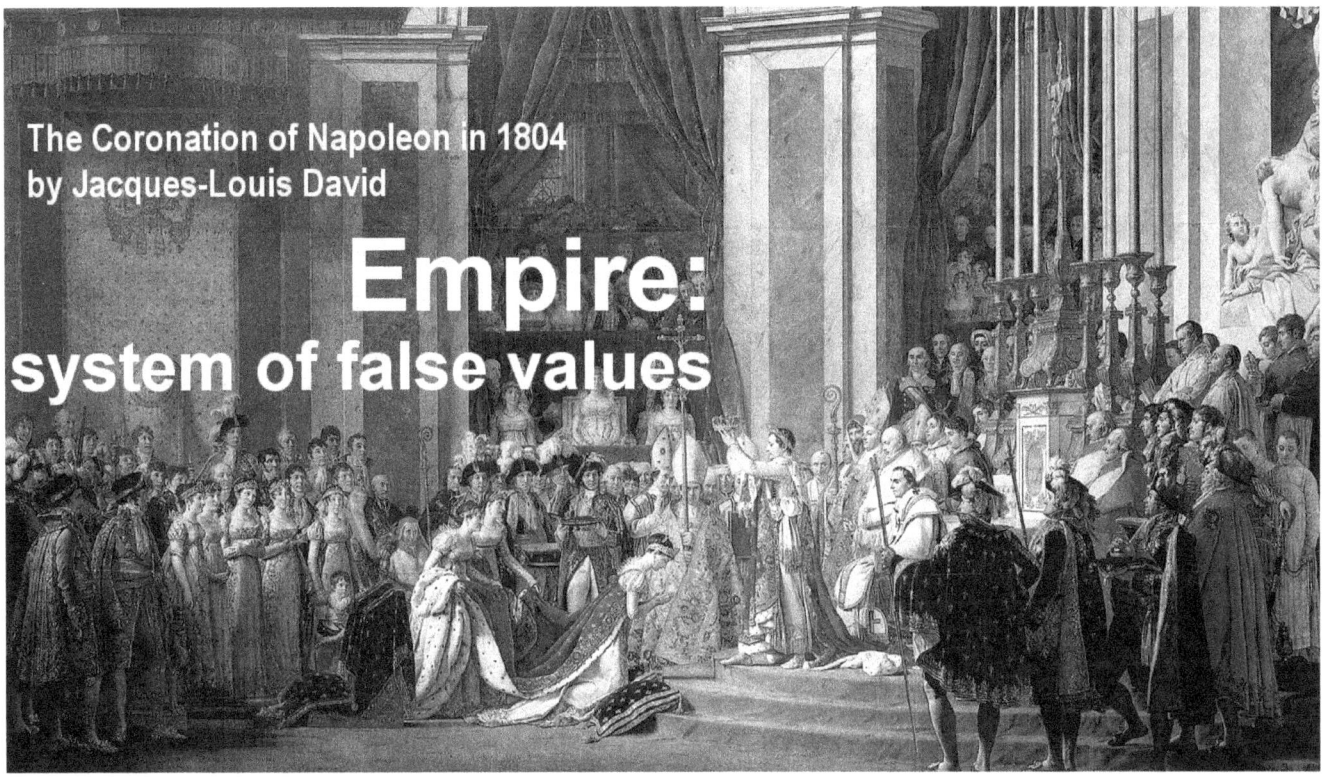

Empire: a system of false values.

Empire is a force directed against humanity by which empire dooms itself

The Coronation of Napoleon in 1804 by Jacques-Louis David

Hiding behind its gilded facade, Empire is a force directed against humanity by which empire dooms itself as the only declared enemy that humanity has. Its real name is war. All wars are by Empire, for Empire, and on behalf of Empire, or against competing Empires. Empire and war is one. Empire is a force deployed to blunder, subdue, and to steal, instead of to create, to produce, and to build. By this, empire itself is doomed.

Empire is also the enemy of science

Empire is also the enemy of science, and the only inherent enemy that science has, and ever will have.

Our astrophysical Sun, is being paraded on posters with a false face

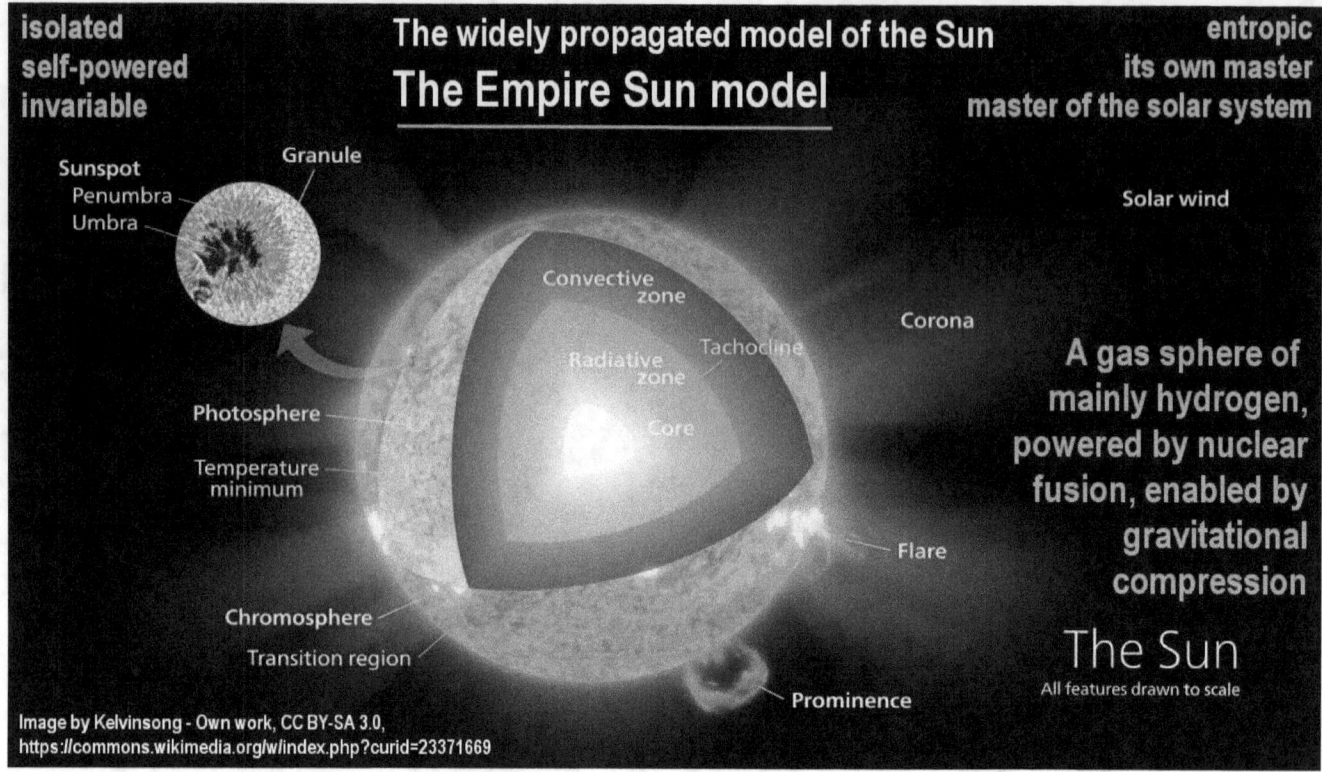

This may be the reason why in modern time, the star for the Earth, our astrophysical Sun, is being paraded on posters with a false face. It is portrayed with a face of which nothing is actually real, but which reflects the nature of Empire through and through.

It would be surprising if this wasn't so, because anything that Empire creates is a paradox onto itself.

But there is light at the end of the tunnel symbolized by the cannon fire

But there is light at the end of the tunnel - a light as symbolized by the cannon fire of the 1812 Overture that appears to celebrates the explosion of truth over what would suppress it.

The Empire Sun poster is strung across the landscape of science

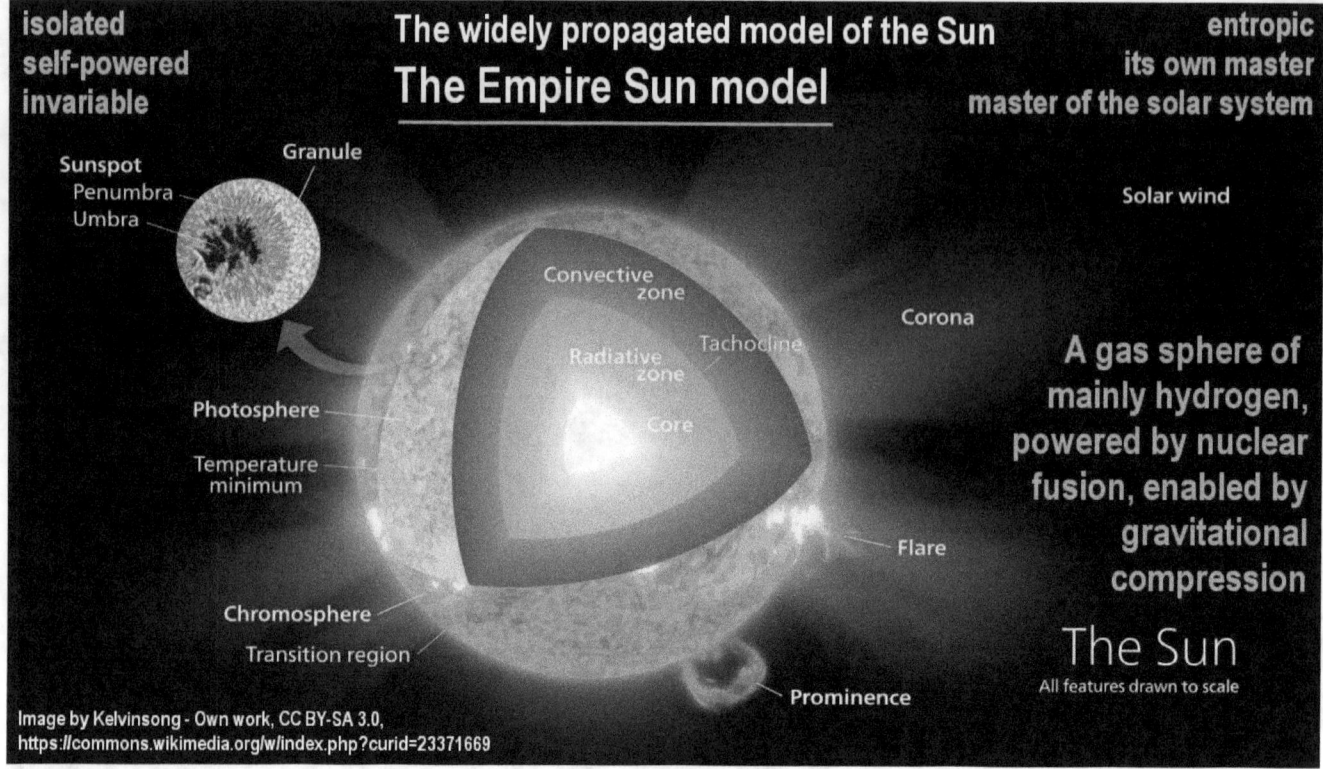

The Empire Sun poster is strung across the landscape of science. It parades the Sun in the form of a sphere of hydrogen gas that is self-powered by nuclear fusion reactions occurring at its core. This Sun is deemed to exist disconnected from the universe - self-made, as its own master, burning its own fuel, subject to nothing, invariable, unchanging, and forever constant in its operation.

Isn't this also what empire says about itself, in essence?

Empire has devised a model for the Sun that mirrors its own image

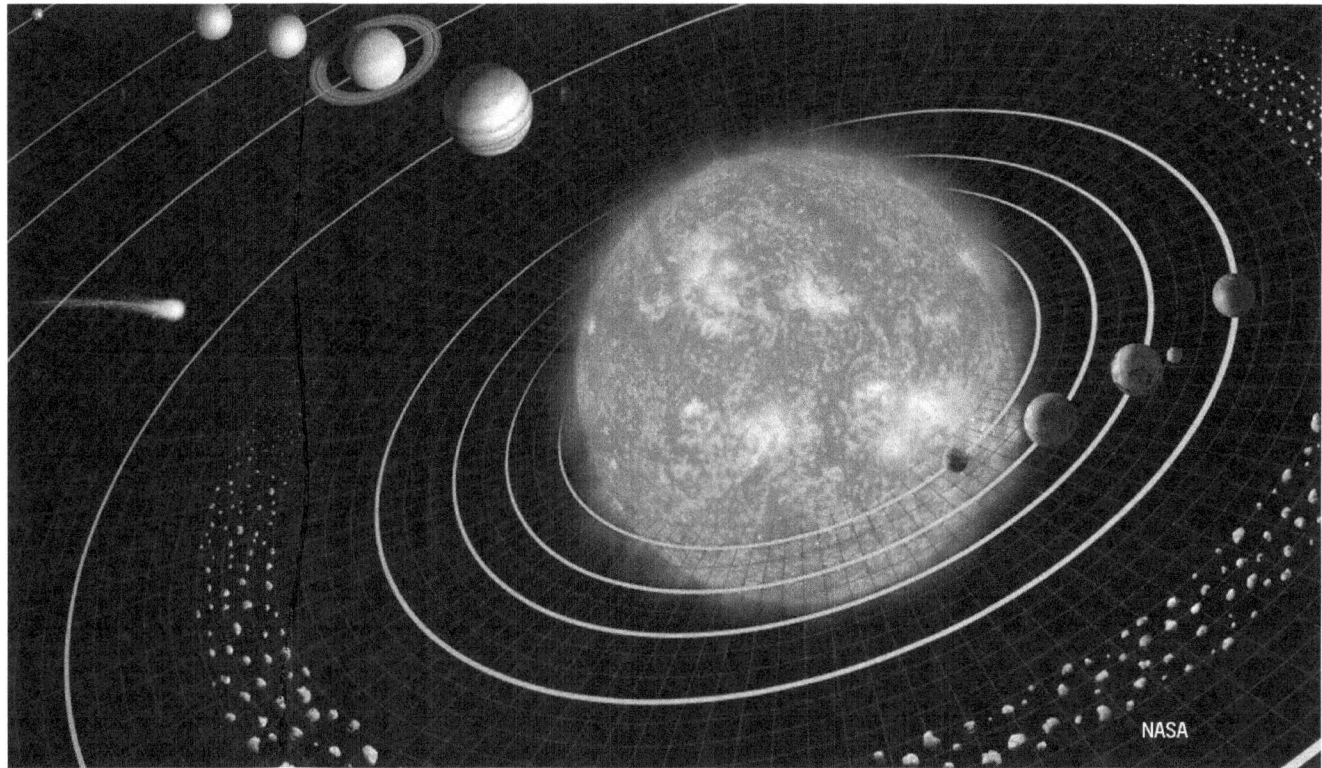

It is not surprising in this context that empire has devised a model for the Sun that mirrors its own image, a kind of astrophysical model that reflects the nature of empire. The model that resulted isn't likely a conspiracy. It is most likely honestly derived.

Empire paints the world with the colors of its ideology

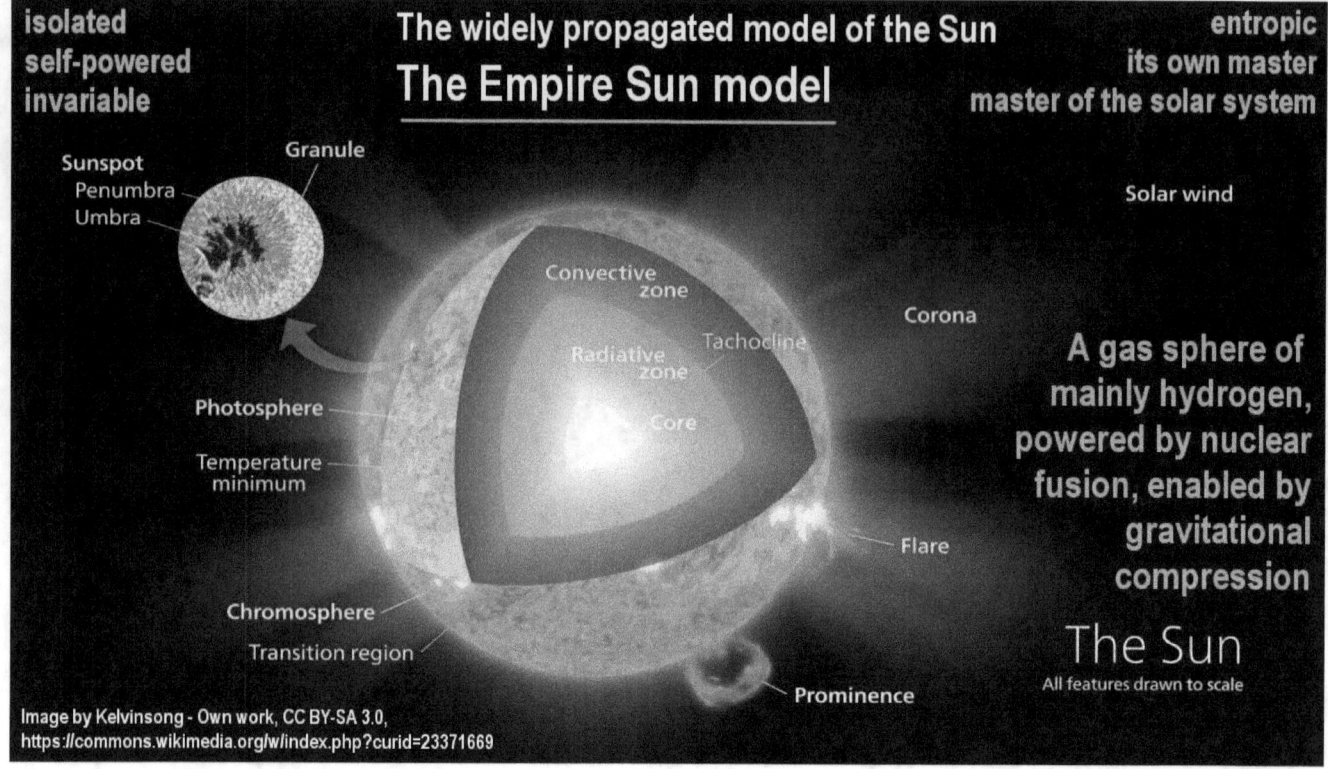

In fact it would be surprising if the Empire-created an image of the Sun that did not reflect in principle the characteristics of the system of empire. Empire paints the world with the colors of its ideology.

Every great artist paints the images of the world as they appear in the artist's mind

Every great artist paints the images of the world as they appear in the artist's mind. In this context, the paintings of Leonardo da Vinci and Pablo Picasso portray different views of the same object, seen through different eyes, shaped by different ideals. It would be surprising if this differentiation in characteristics didn't exist. This applies to science likewise.

The masters of empire who control the sciences are steeped in the ideology they serve

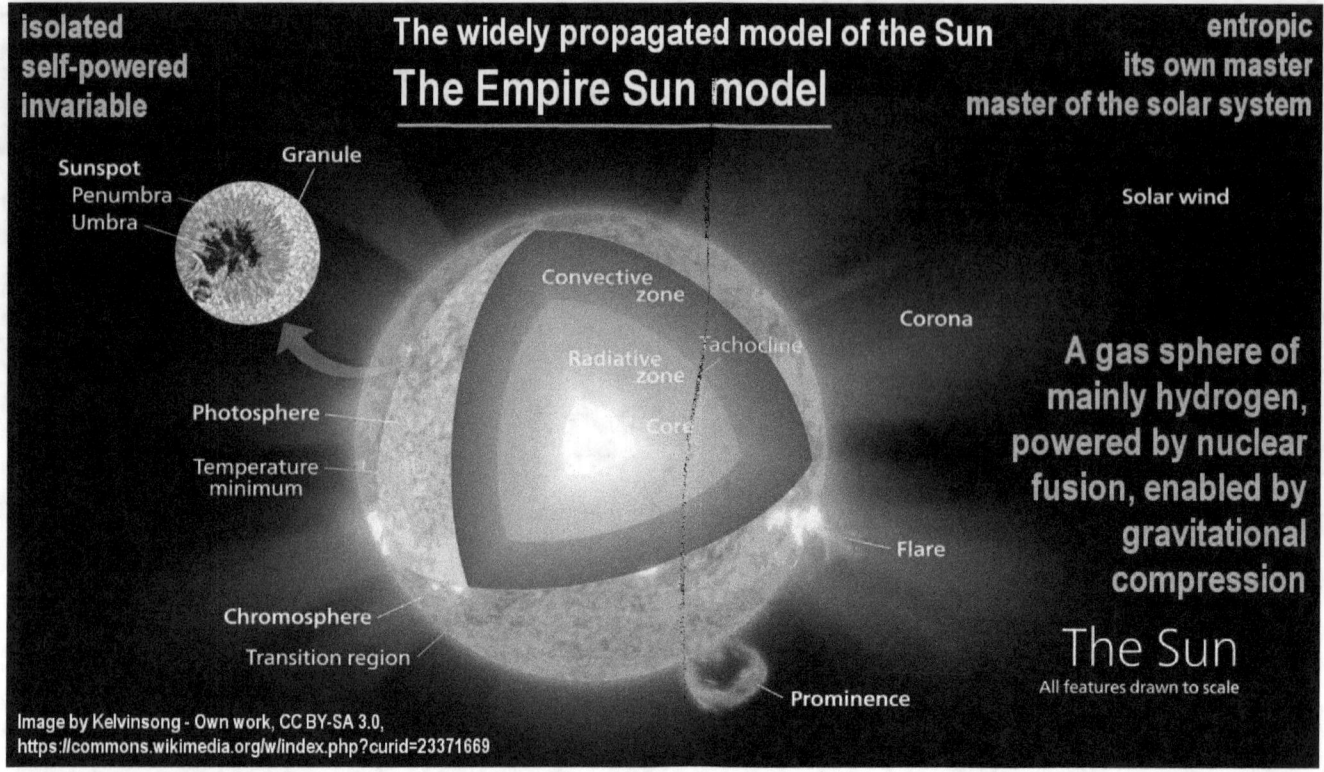

The masters of empire who control the sciences, just as they control nations subversively, are invariably steeped in the ideology of empire that they serve. As a consequence, the empire-created poster of the Sun and its operation bears the reflected image of the ideology of empire. It serves a specific objective instead of the truth. The disorientation has consequences of course - sometimes extreme consequences. Some of these are intentional.

Under the Empire-Sun model - and I mean "empire" in the generic sense as a system of thinking that stands poised in the modern world to repeat the Napoleonic plunder of 1812, though on a vastly more-- gigantic scale this time around, humanity is enticed into a type of dreaming in which the Next Ice Age is denied, even while it stands deadly poised and as close as the 2050s. The induced dreaming that the Ice Age is neither real nor near, is dangerous, because the sleep state blocks all efforts for building the necessary infrastructures in preparation for the coming conditions.

Living oblivious of the truth the Ice Age is being swept under the rug

Living oblivious of the truth, under the 'flag' of the Empire Sun, the greatest danger to society's existence as the Ice Age looms before us, is being swept under the rug. It was the same for the soldiers who were crossing the river Niemen into Russia. They were oblivious of the doom they were led into under the banner of Empire that projects a myth of glory that is fundamentally a lie.

Oblivious humanity is poised to re-experience the doom of this folly

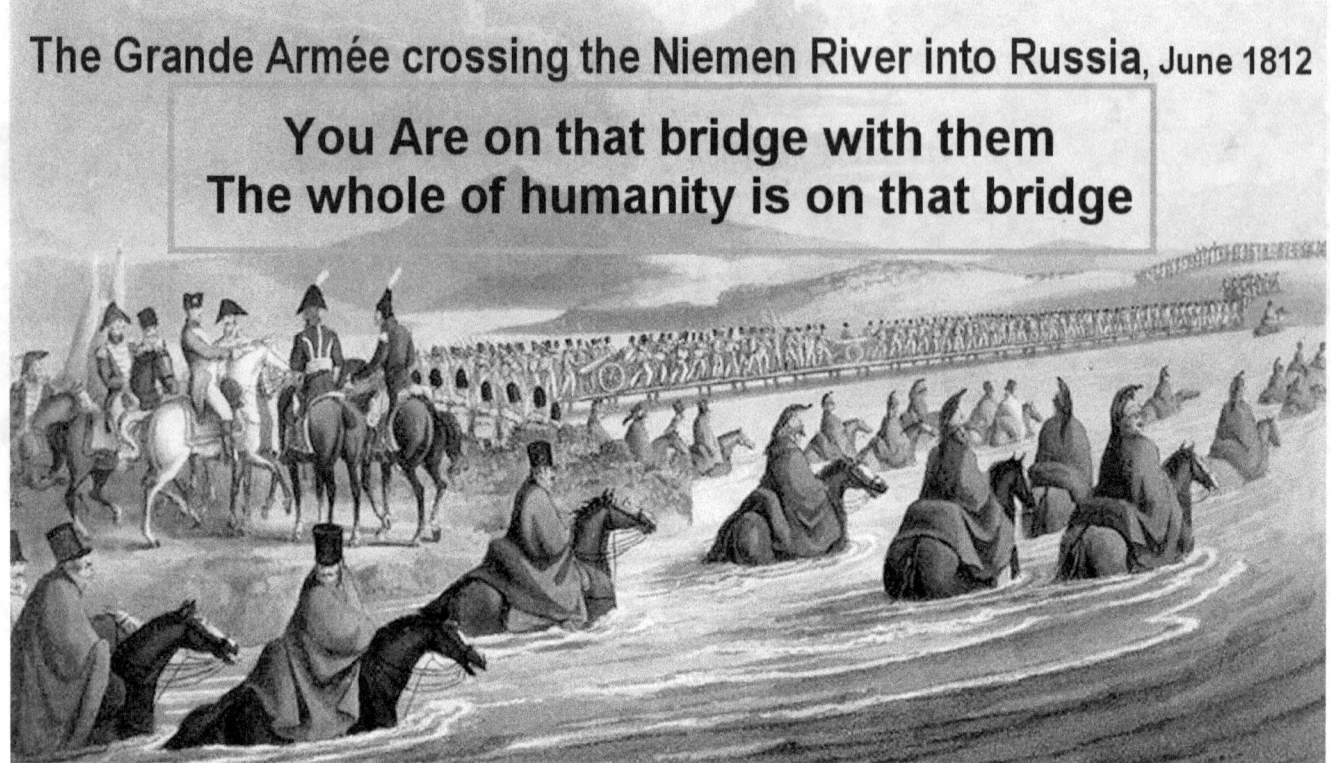

Oblivious in the same sense, humanity is poised to re-experience the doom of this folly in the near future; in a world without food, scorched landscapes, and evermore unbearable drought and cold.

That this fate is near, is evident by the fact that the infrastructures are not being built

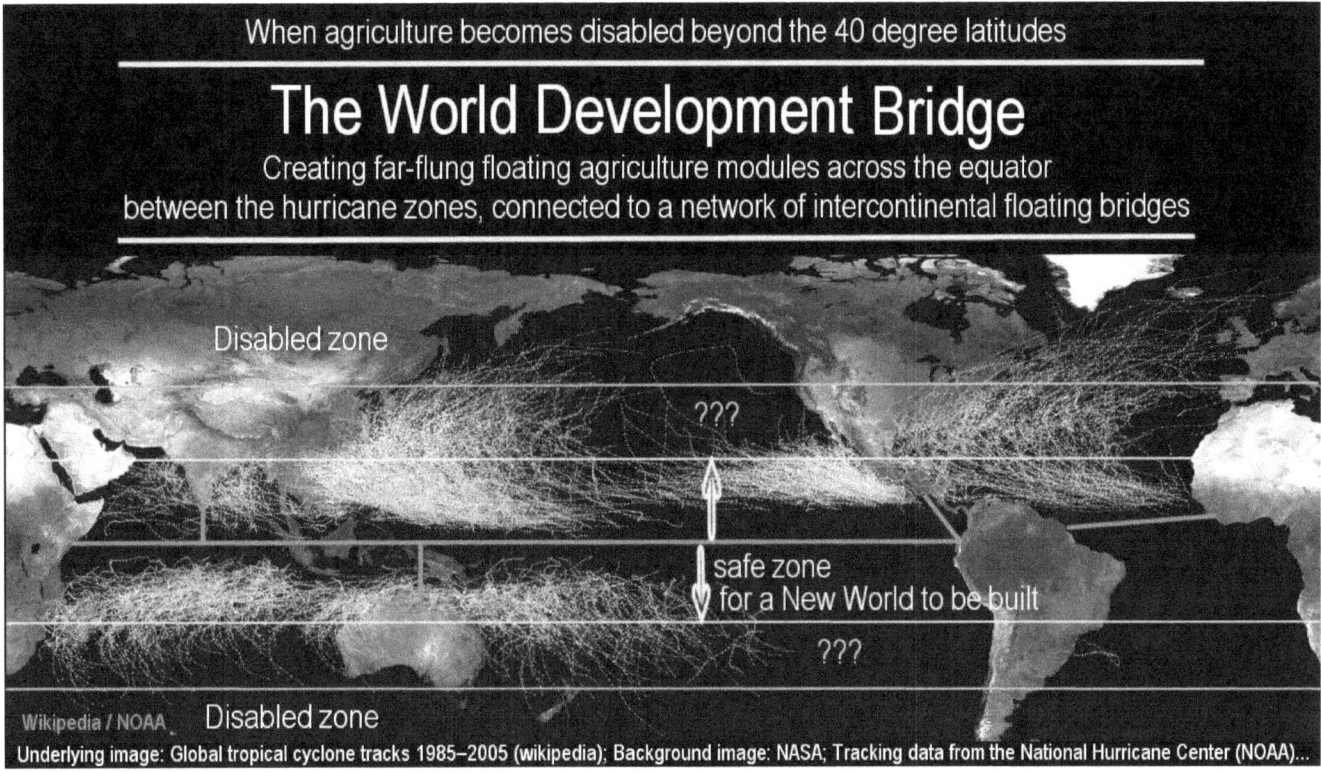

That this fate is near, is evident by the fact that the infrastructures are not being built that enable agriculture and human living to continue in an Ice Age environment.

We have a chance today, in our time, to avoid this doom - a chance that the soldiers on the bridge did not have.

➤ Light at the end of the tunnel

Light at the end of the tunnel

(beyond the Empire Sun)

Light at the end of the tunnel - beyond the Empire Sun.

The Real Sun is different from the Empire Sun in every respect

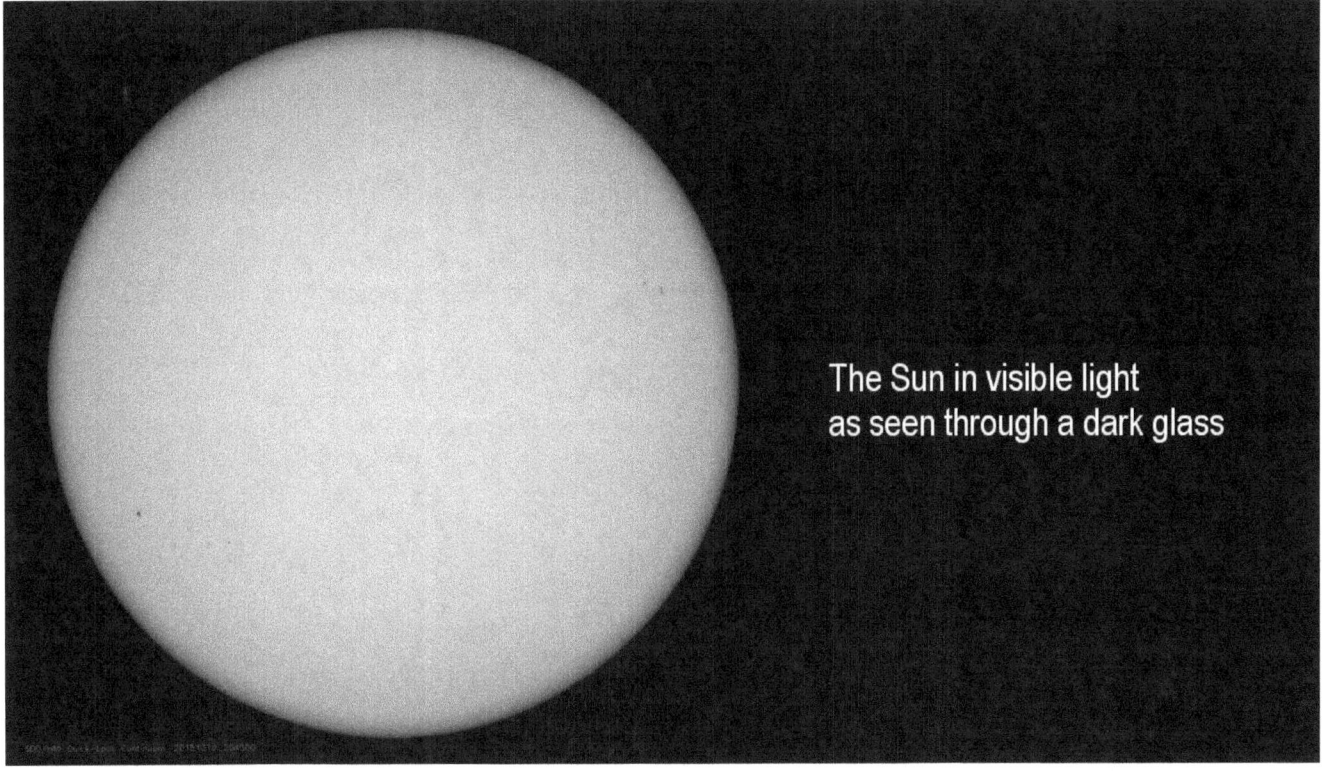

The Sun in visible light as seen through a dark glass

We have a chance today to discern the Real Sun. The Real Sun is different from the Empire Sun in every respect.

The Real Sun is vulnerable; variable; a sphere of plasma that is largely empty inside and is energized externally by principles that render it as an integrated element of an anti-entropic universe.

On its surface, the largely empty Sun is teeming with energetic electric interactions.

On its surface, the largely empty Sun is teeming with energetic electric interactions. This turbulent sea of electric currents flowing in plasma is not of the Sun's own creating. It merely facilitates it, as a catalyst for it.

Wherever electric movements happen, magnetic phenomena follow

Also, wherever electric movements happen, magnetic phenomena follow. All magnetic phenomena are electric effects - they are the effect of electric particles in motion. The Sun is covered with them, even while it is empty inside.

But what proves the seeming 'paradox?' What proves that our mighty Sun is empty inside, so that what we of it, is but skin-deep?

One of the Sun's smallest features closes the door on the Empire Sun model

The proof of the seeming paradox is located in one of the Sun's smallest features that all by itself closes the door on the Empire Sun model. This feature is the sunspot.

The sunspot is a dark spot on the face of the Sun. It results when a hole is ripped into the shiny surface of the Sun.

Whether the sunspots are small in size or large, all share one common characteristic

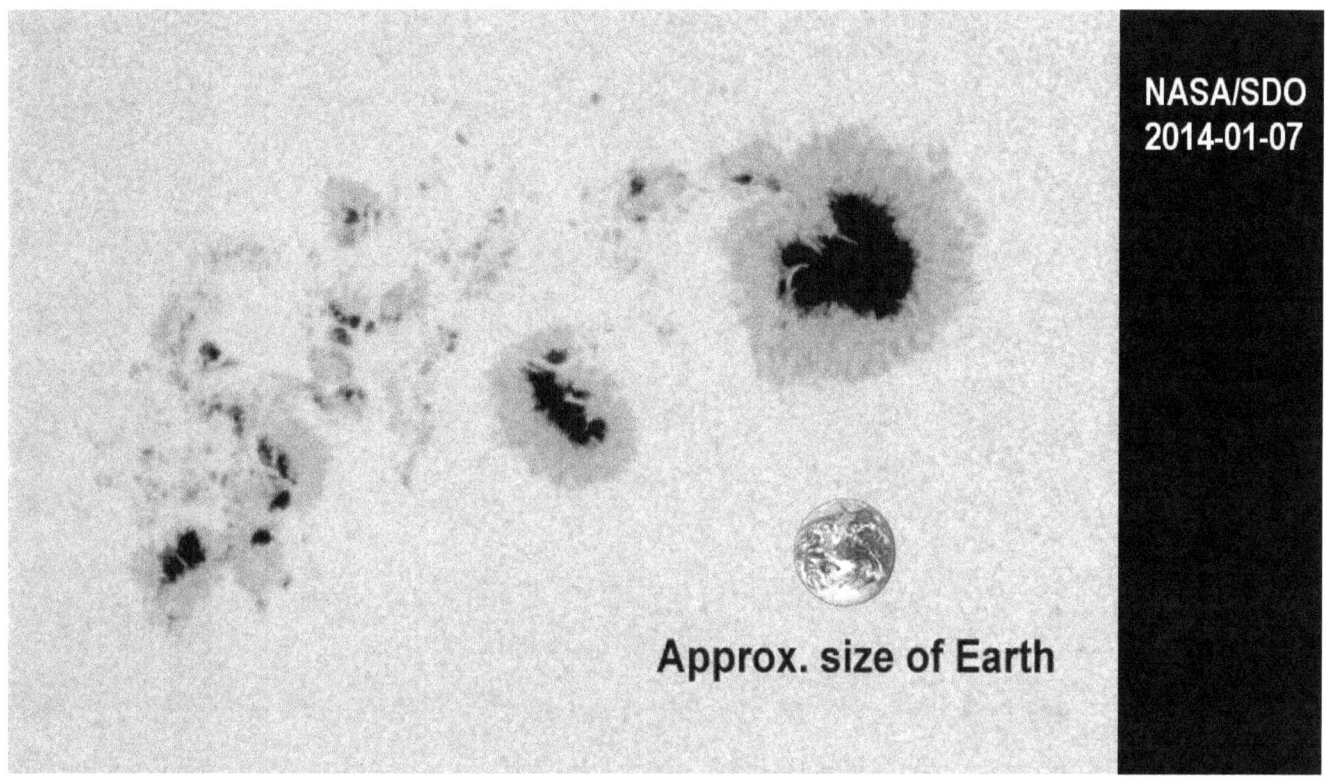

Whether the sunspots are small in size or large, or many or few, they all share one common characteristic. Their umbra are dark, because nothing exists below the shiny surface.

No matter how closely we look, and how much we expand the view

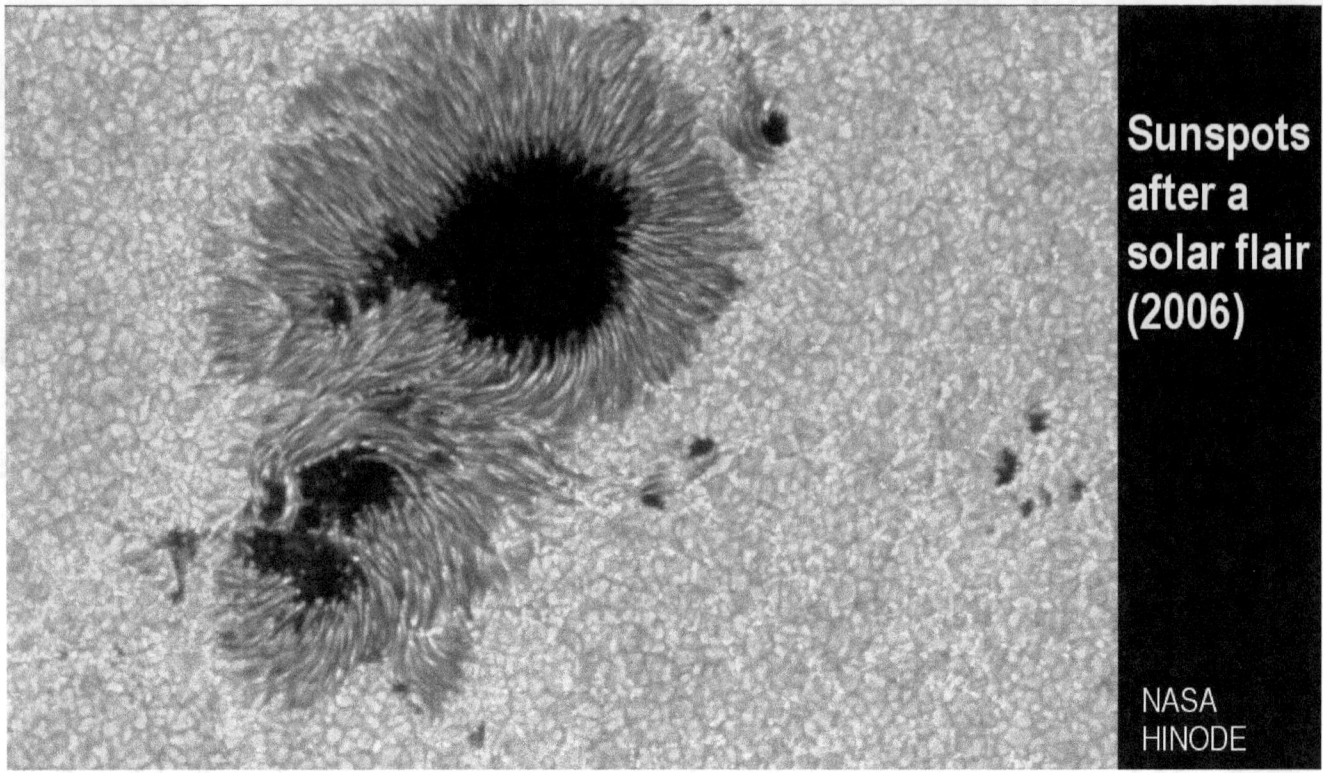

No matter how closely we look, and how much we expand the view, we always look into a void when we look through the umbra of a sunspot.

Paradoxically, what we behold in the void, is a key feature of the very existence of the universe.

If one was given the task to design the perfect universe, created out of nothing but energy and principles, one would end up with the kind of design that we actually have before us.

Throughout the universe, 99.999% of all mass exists in the form of plasma

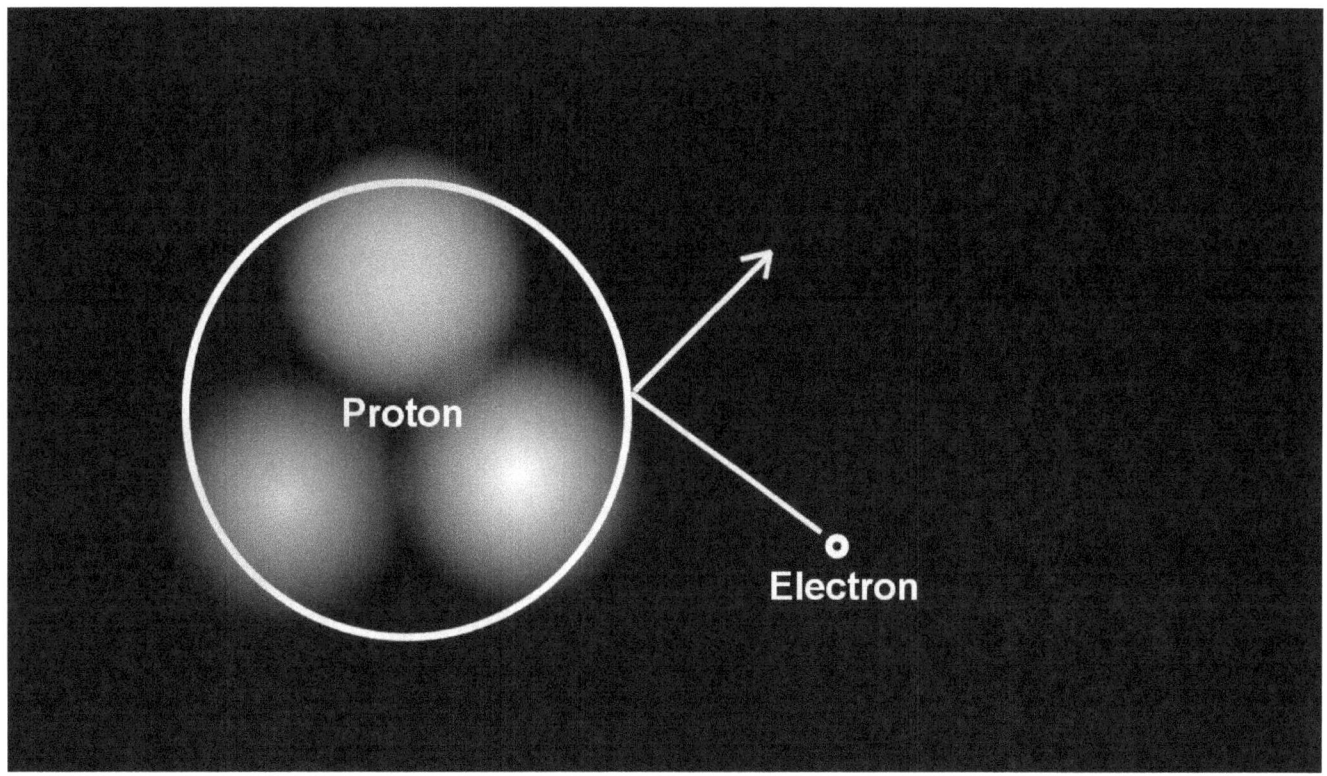

Throughout the universe, 99.999% of all mass exists in the form of plasma. Plasma is the term for the basic building blocks of the universe, termed protons and electrons, existing in unbound form.

The electrons and protons themselves are recognized to be constructs of quarks, of up-quarks and down-quarks, and so on.

Quarks are understood as but moving points of energy

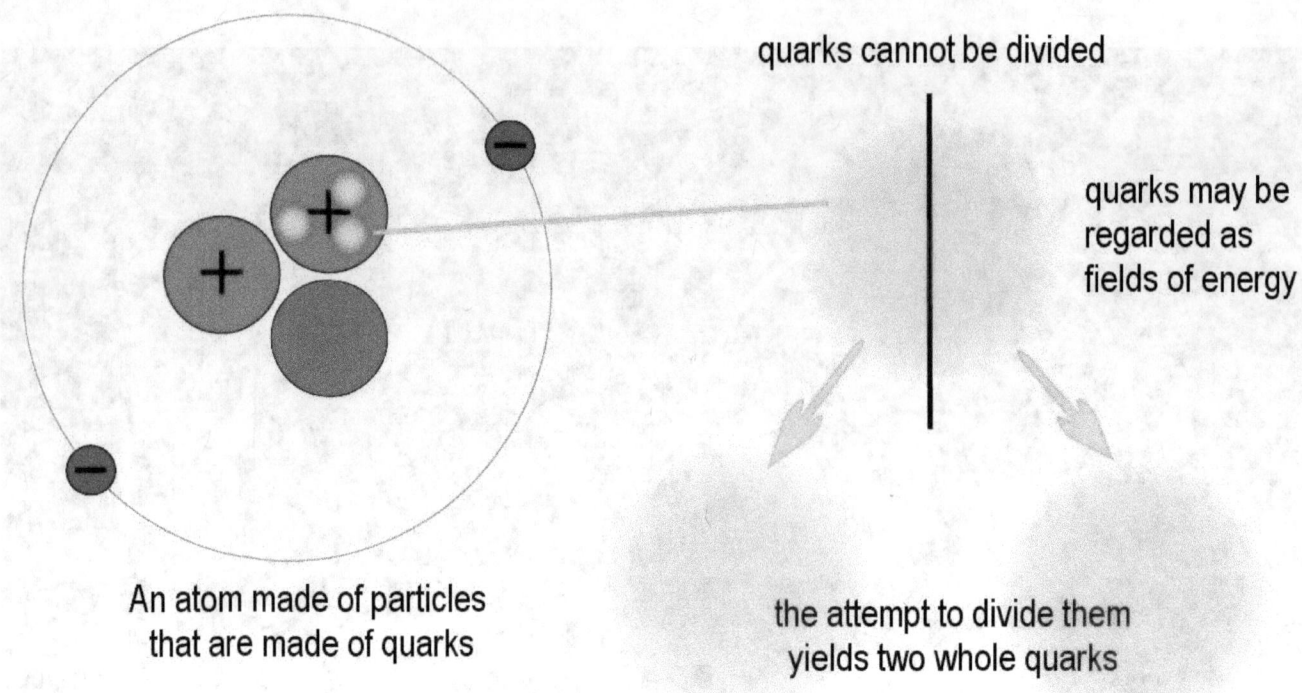

Quarks are understood as but moving points of energy. This means that all plasma, and all atoms in the universe are derived from them.

Explicate features of the implicate order of the energy background of cosmic space

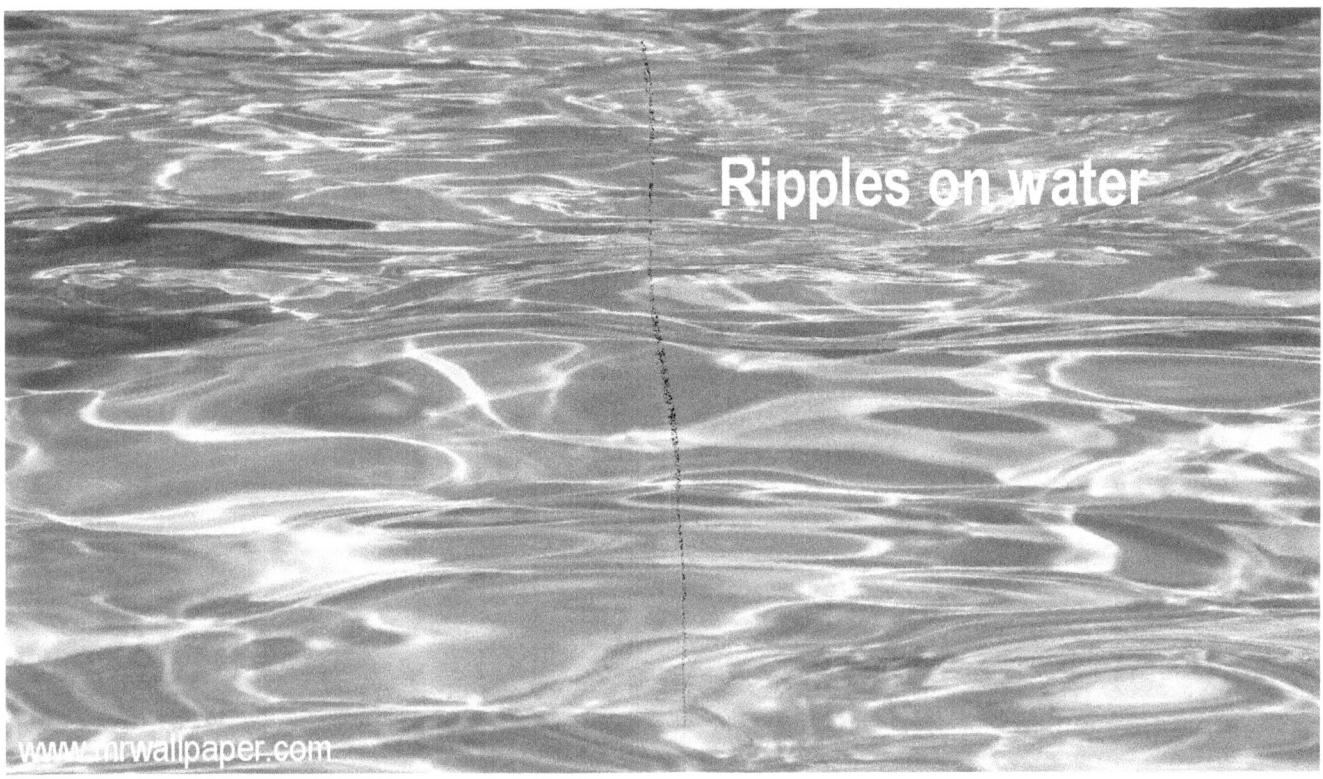

And the quarks themselves are but explicate features of the implicate order of the energy background of cosmic space.

That's how the great theoretical physicist David Bohm had perceived the universe

That's how the great theoretical physicist David Bohm had perceived the universe, whom Einstein once referred to as his successor.

The specific arrangements of the quarks are recognized

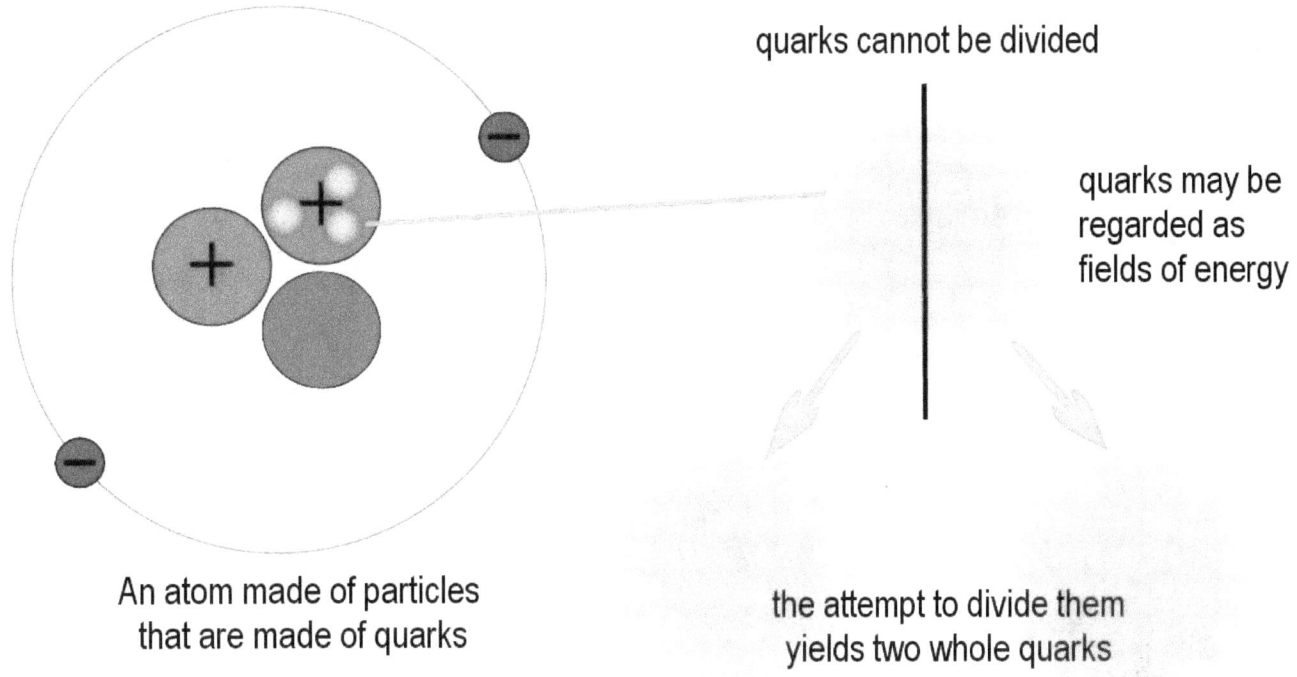

An atom made of particles that are made of quarks

quarks cannot be divided

quarks may be regarded as fields of energy

the attempt to divide them yields two whole quarks

The specific arrangements of the quarks are recognized to give the electrons and protons their electric potential - positive for the proton, and negative for the electron.

With the same force, particles with equal charge repel one another

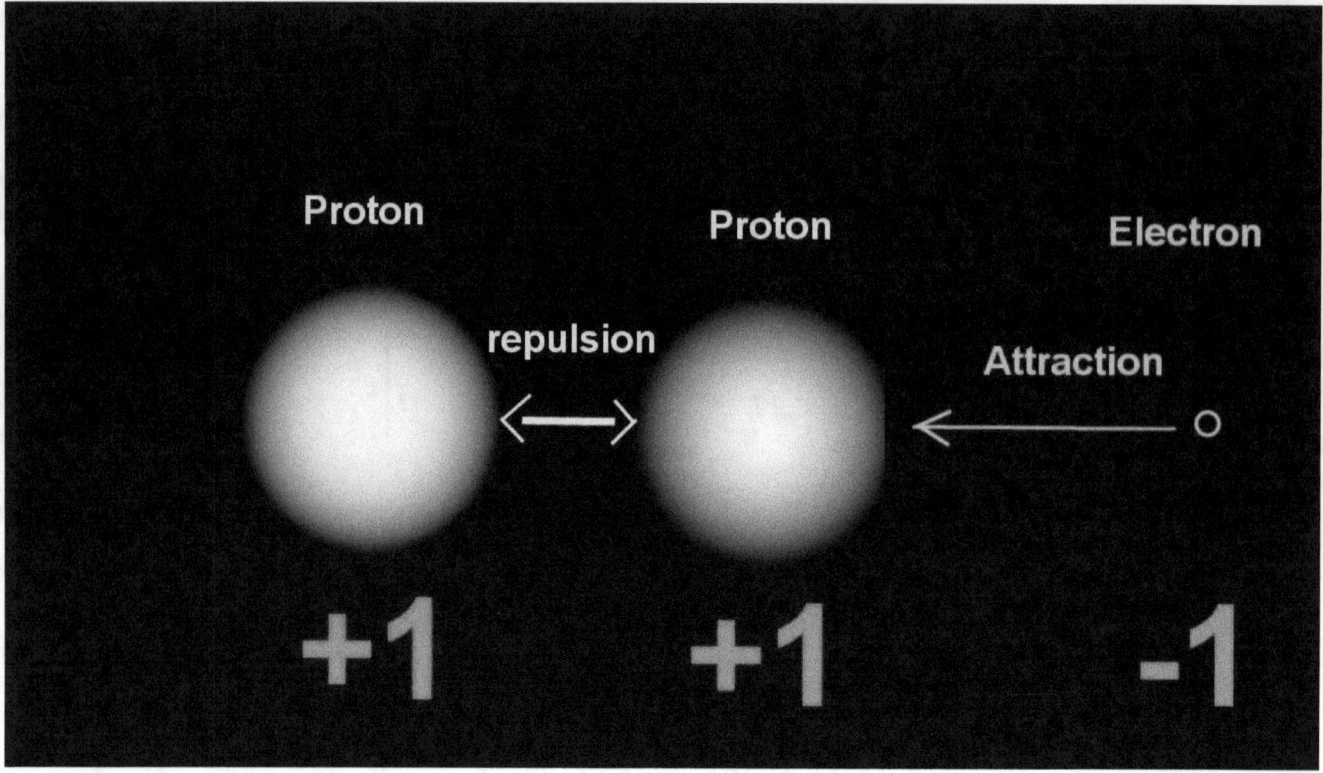

By the resulting electric potential the unlike plasma particles, by their complementary nature, attract one another with the electric force of the universe.

With the same force, particles with equal charge repel one another, which creates more room around them.

At very close distances, a strong nuclear force repels the two apart

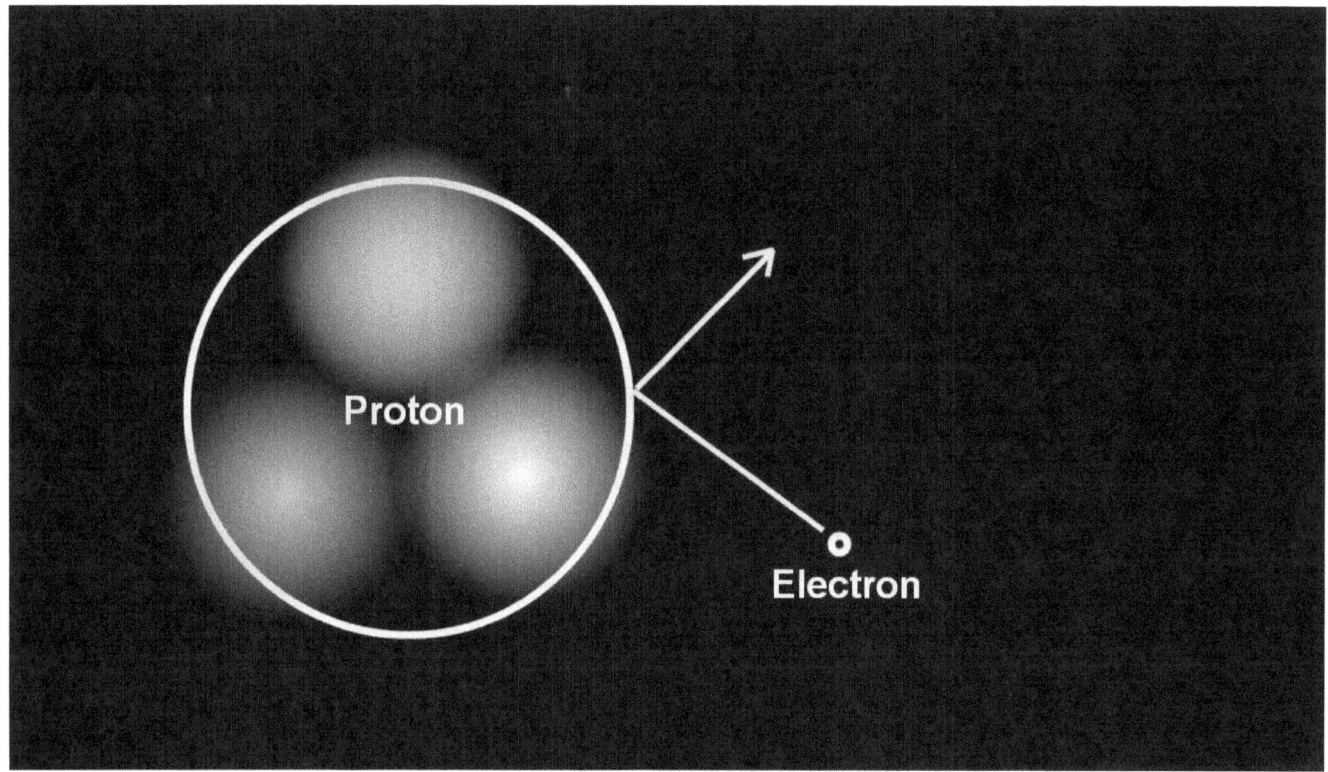

However, the universe doesn't allow the complementary attraction to go so far that the attracting particles latch on to each other. At very close distances, a strong nuclear force from within the combination repels the two apart. Without that we wouldn't have a dynamic universe. After the electrons bounce away, they are promptly attracted again by the electric force. The end-result is that the electrons in plasma, which are 2,000 times smaller and lighter than the protons are, are drawn into an endless dynamic dance of attraction and repulsion, around the protons.

On this basis plasma becomes bound up into atoms

**atoms are formed by the dynamic 'dance'
of electrons being attracted and forced to rebound**

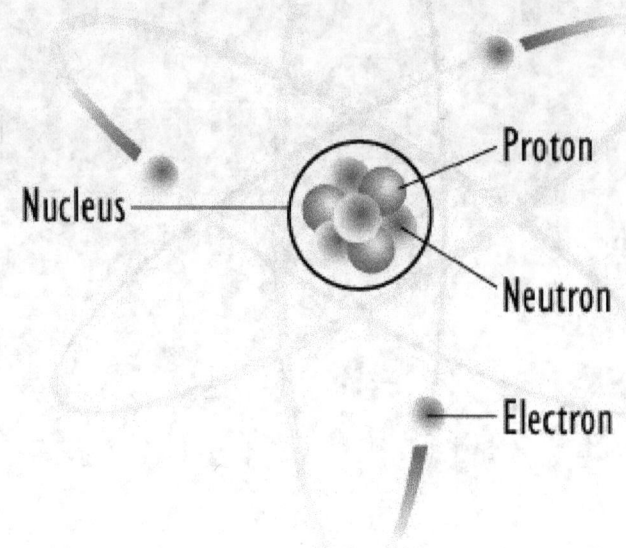

wikipedia (image)

On this basis plasma becomes bound up into atoms that subsequently form the planets and everything on them.

When plasma exists in unbound form in dense concentration

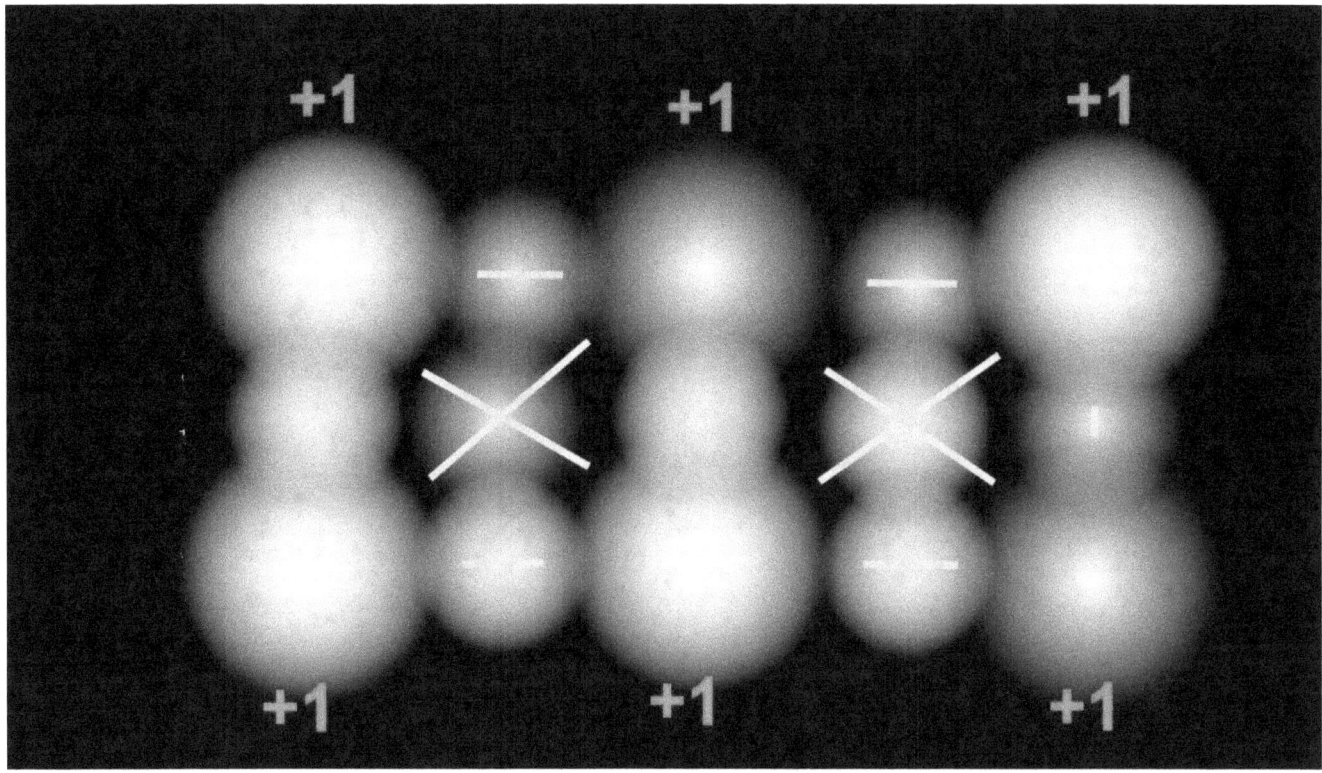

When plasma exists in unbound form in dense concentration, the free dancing of the electrons forms an electric barrier between the free protons. The barrier inhibits the protons repelling one-another. Paradoxically, this insignificant-seeming feature has a world-shaping effect.

The result is that extremely dense concentrations of mass can be forged that way, with a greater density than atomic packages can have, by the electrons shielding the protons against their repelling force. The brown dwarf star is an example.

Brown dwarf stars are typically of the size of Jupiter

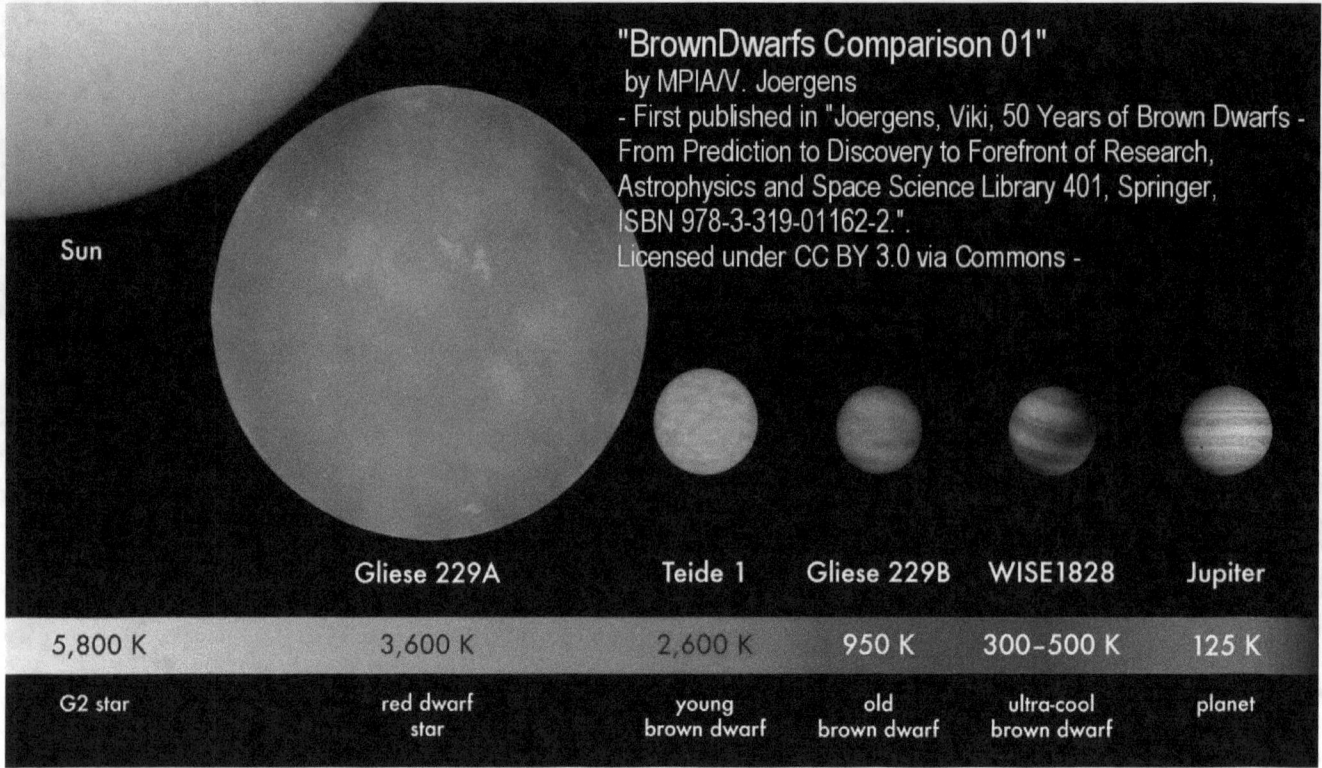

Brown dwarf stars are typically of the size of Jupiter. However, they can contain up to 75 times as much mass as Jupiter does. This makes them the densest objects that we know of, with an up to 15-times greater density than the Earth.

Extremely large concentrations of plasma, gravity begins to play a role too

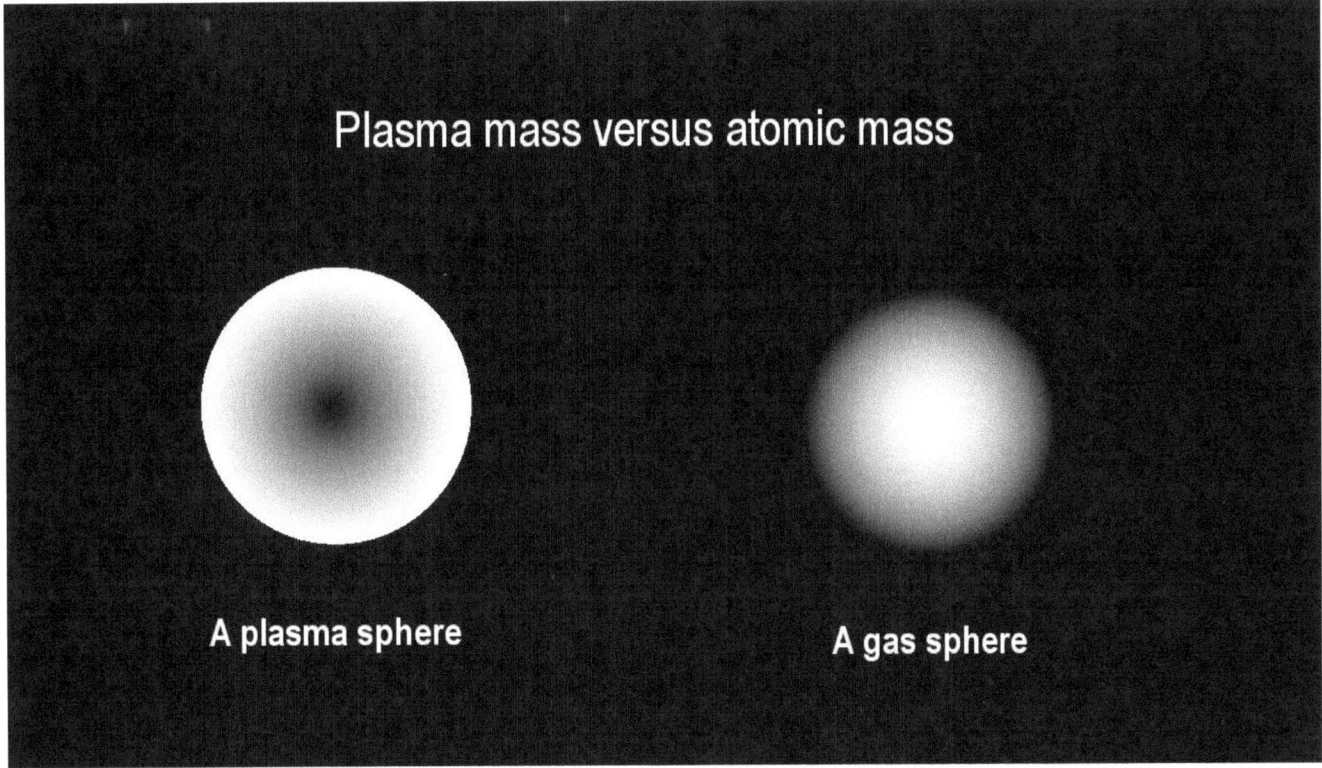

However, in extremely large concentrations of plasma, gravity begins to play a role too. With the protons being nearly 2,000 times heavier than the electrons, the lighter electrons tend to migrate to the surface, like marshmallows do on milk.

➤ **A sun must be 'hollow' to function**

A sun must be 'hollow'

(to function)

A sun must be 'hollow' to function

The migration of the electrons away from the center of gravity

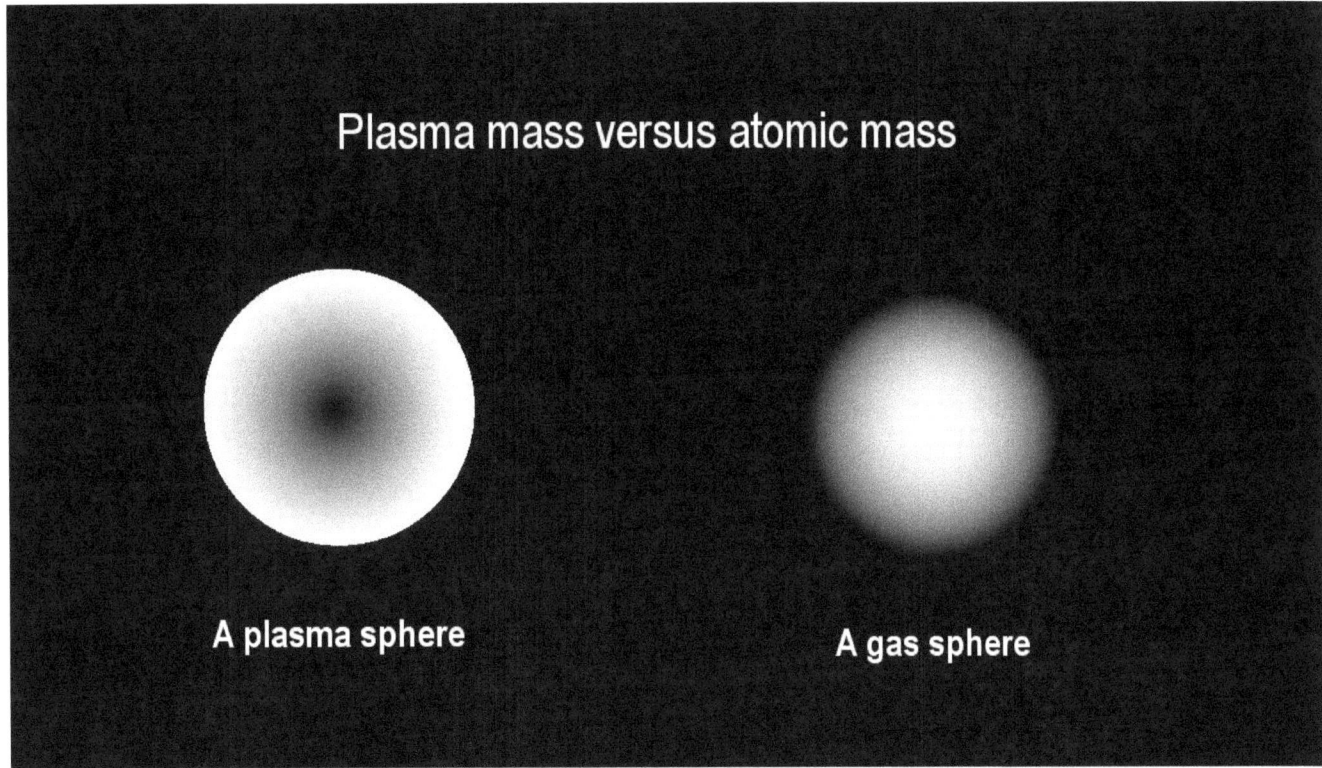

The migration of the electrons away from the center of gravity, enables the needed high-electron density on the surface of a sun that is required for a sun to function. It also enables the protons in the core of a sun to see each other more strongly, and thereby repel each other more strongly, whereby a very- large plasma sphere results has the lowest mass density at its core, and the highest density at its surface where the greatest electron density ends up, which makes the resulting sun an extremely reactive electric catalyst.

This principle of electron migration is evidently one of the most-critical features of the universe, because it enables extremely large stars to exist with an extremely low over-all mass density. The resulting inverse ratio enables a sun to have the kind of extremely large surface area that would otherwise not be possible. A sun must have a large surface area in order to be able to radiate large volumes of energy.

The principle defines the apparent paradox of our Sun being an extremely radiant star, and at the same time to be an almost empty star.

A sphere of atomic gas, in contrast, lacks the electric features of plasma

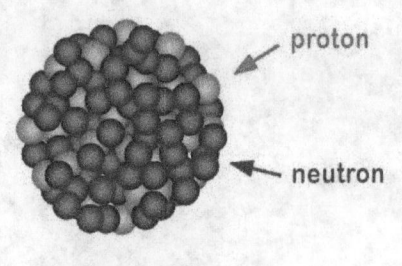

Uranium Nucleus
92 protons, 146 neutrons

Atomic construction principles

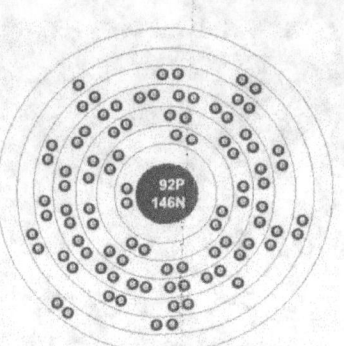

92 electrons, in 7 shells of orbital spaces

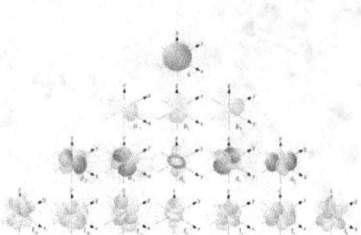

A shell can contain up to 6 levels of subshells

Source:
http://www.saburchill.com/chemistry/
http://www.ganil-spiral2.eu/science-us/

A sphere of atomic gas, in contrast, lacks the electric features of plasma. Atoms are electrically neutral structures. They have no external electric effects. They have their constituent plasma particles tightly bound up into electrically balanced, electrically neutral, units.

For this reason, plasma in the form of atoms, can be packed together more densely.

A sphere of atomic elements has its greatest mass density at its core

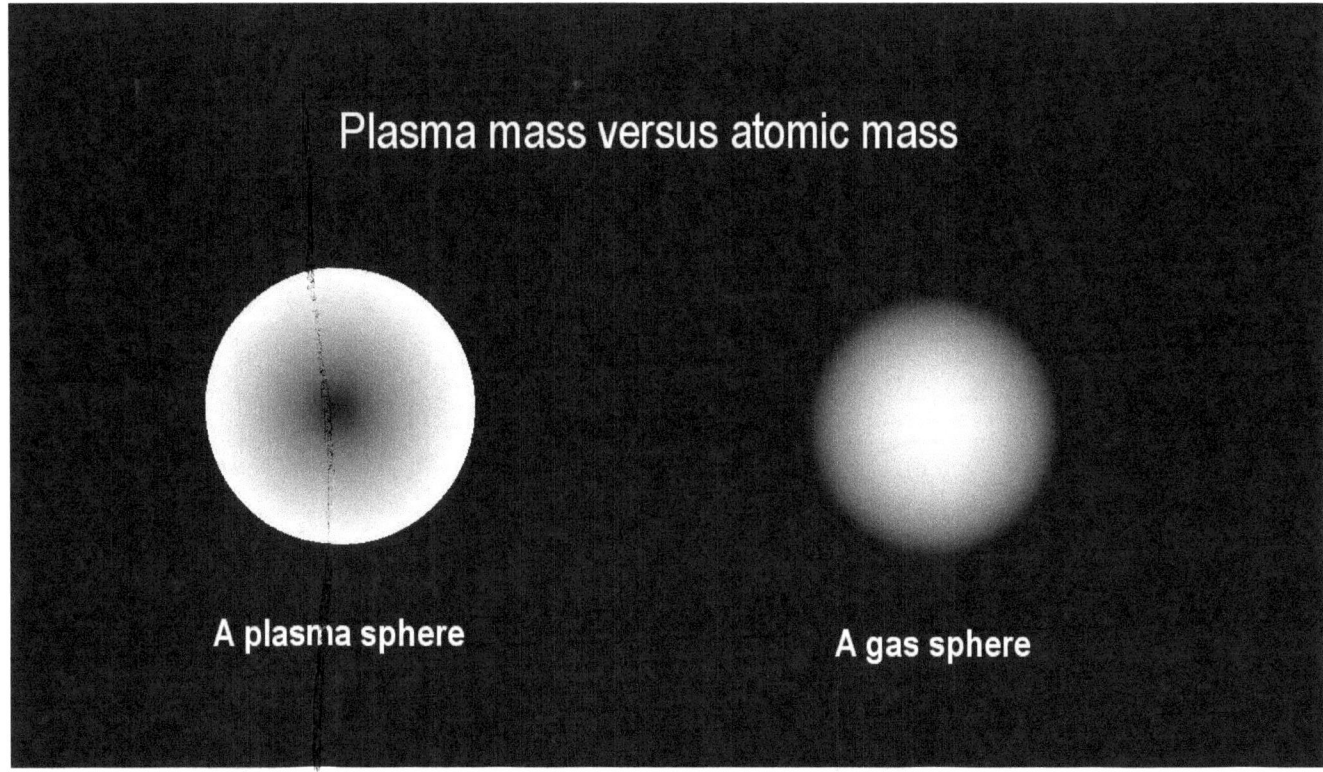

This means that a sphere of atomic elements where gravity is the dominant force, like that of a planet, or of a gas planet, always has its greatest mass density at its core, by gravitational attraction, and the least density at its surface.

The resulting density gradient renders a gas sphere unsuitable for being a sun

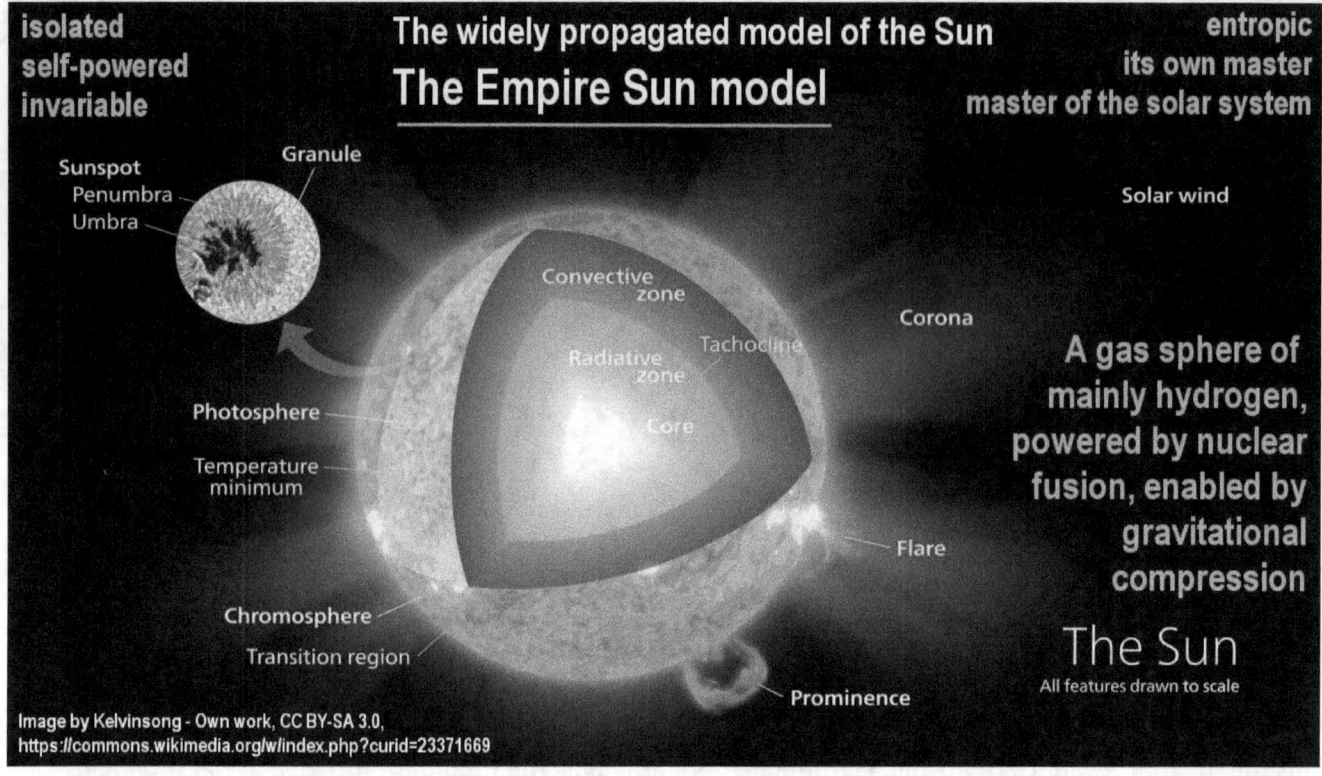

Paradoxically, the resulting density gradient renders a gas sphere unsuitable for being a sun. It actually renders the Empire Sun, physically impossible.

A gas sphere is limited in size by the maximum possible gas compression without the inner atoms being crushed. This limit is overruled in the magical land of the electron degeneracy theory.

Outside the land of magic, the compression limit renders a hydrogen gas sphere too small to ever be a sun. If our Sun was a gas sphere of equal mass, it would be merely a tenth the size it is, if such a dense sphere of gas could actually exist.

In comparing Jupiter with Saturn, Jupiter, which has double the volume of Saturn

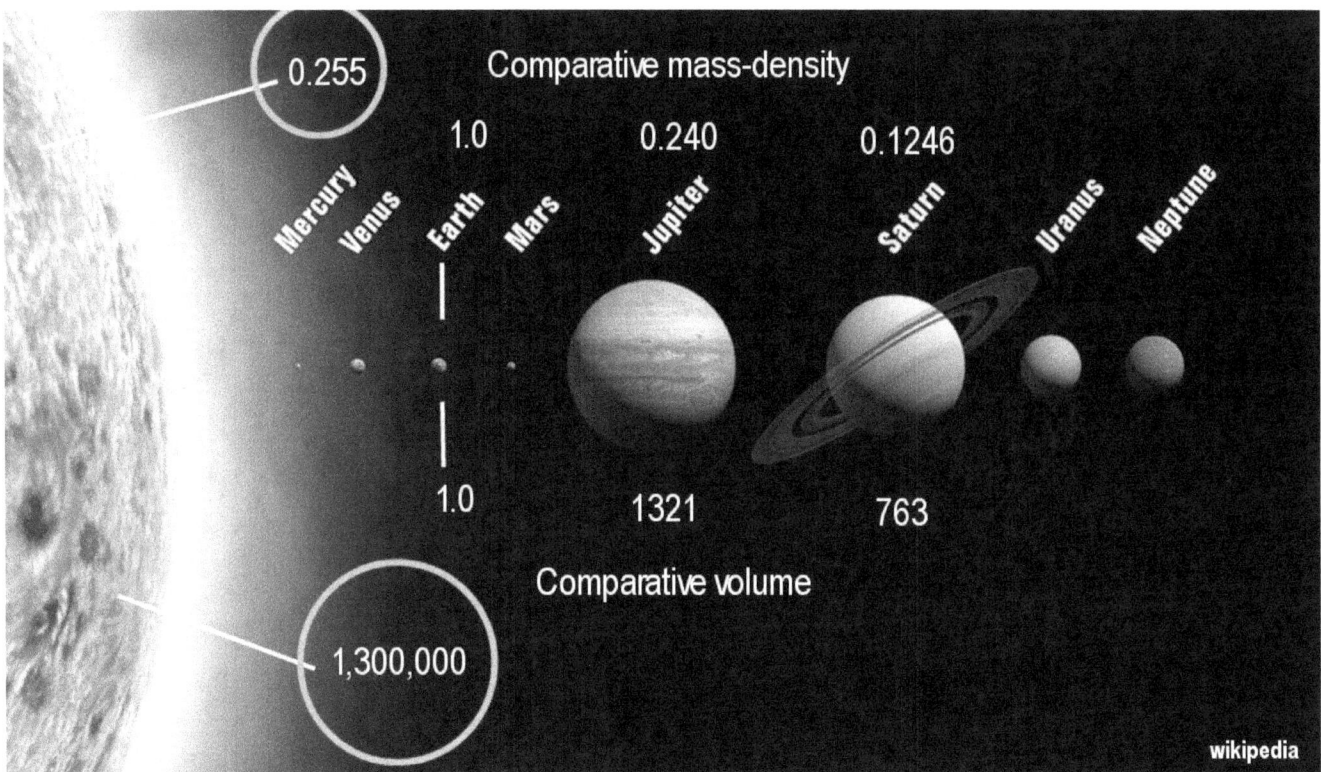

In comparing Jupiter with Saturn, Jupiter, which has double the volume of Saturn, also has double the mass density. The greater mass-density reflects Jupiter's stronger gravity that results in stronger gas compression.

But when one compares Jupiter with the Sun that has a thousand times greater volume than Jupiter, its mass density should likewise be a thousand times greater. But this is not the case. The mass density of the Sun is roughly the same as that of Jupiter. How is this paradox possible outside the box of magic?

The Sun cannot be a gas sphere, but is an empty shell of plasma. with nothing substantial inside of it?

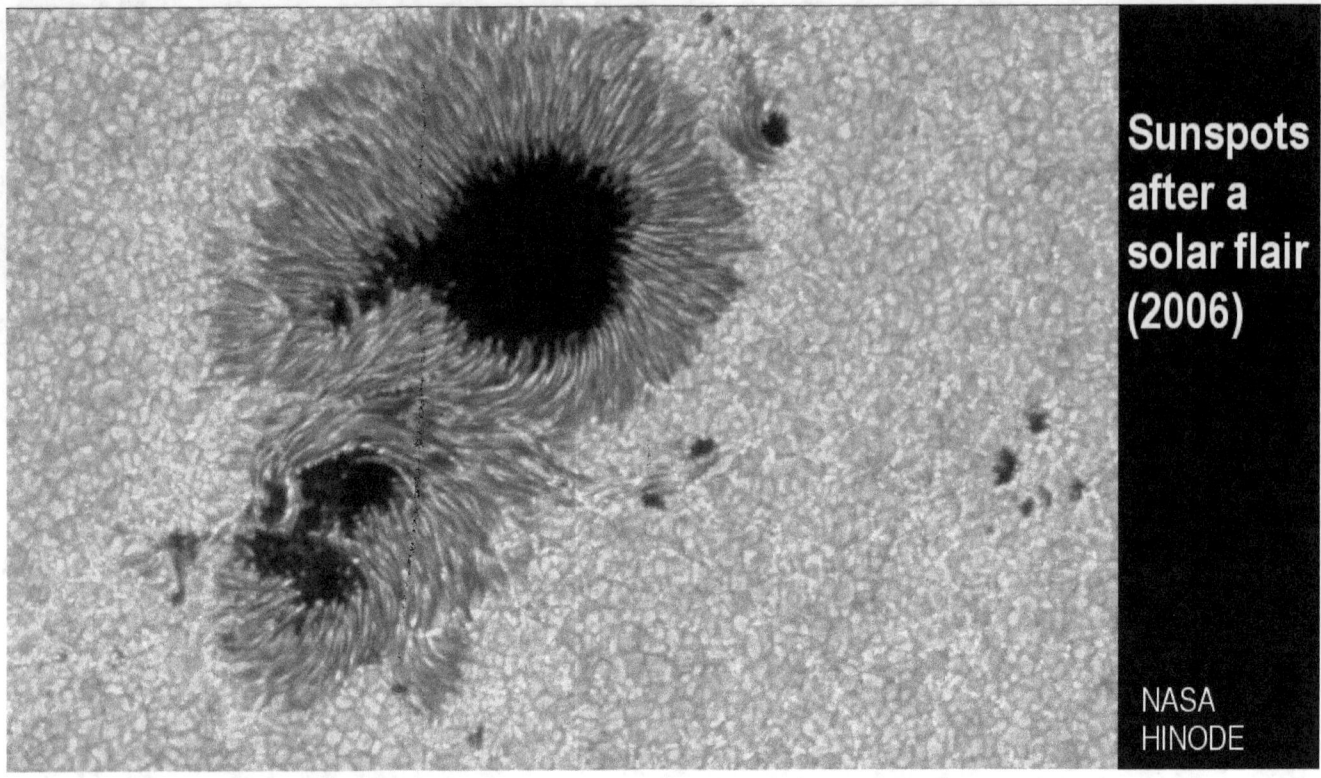

Sunspots after a solar flair (2006)

NASA HINODE

The answer is that the Sun cannot be a gas sphere, but is an empty shell of plasma. with nothing substantial inside of it? Its extremely light weight is justified thereby.

This is also what we see with our own eyes when we look through the umbra of a sunspot on the Sun, which reveals the Sun as dark inside.

Magic would have us see 'dark energy' that blots out light, streaming through the hole in the surface. Without magic, however, we see a dark void, because there is essentially nothing there to be seen below the surface of the Sun.

Indeed why would we see anything different than what we see, with the Sun being an essentially empty sphere that is externally powered? Astrophysical measurements tell us plainly that our Sun cannot be anything else than a largely empty shell of plasma, like an electrically inflated balloon.

If we consider the extremely large star UY Scuty, the largest star

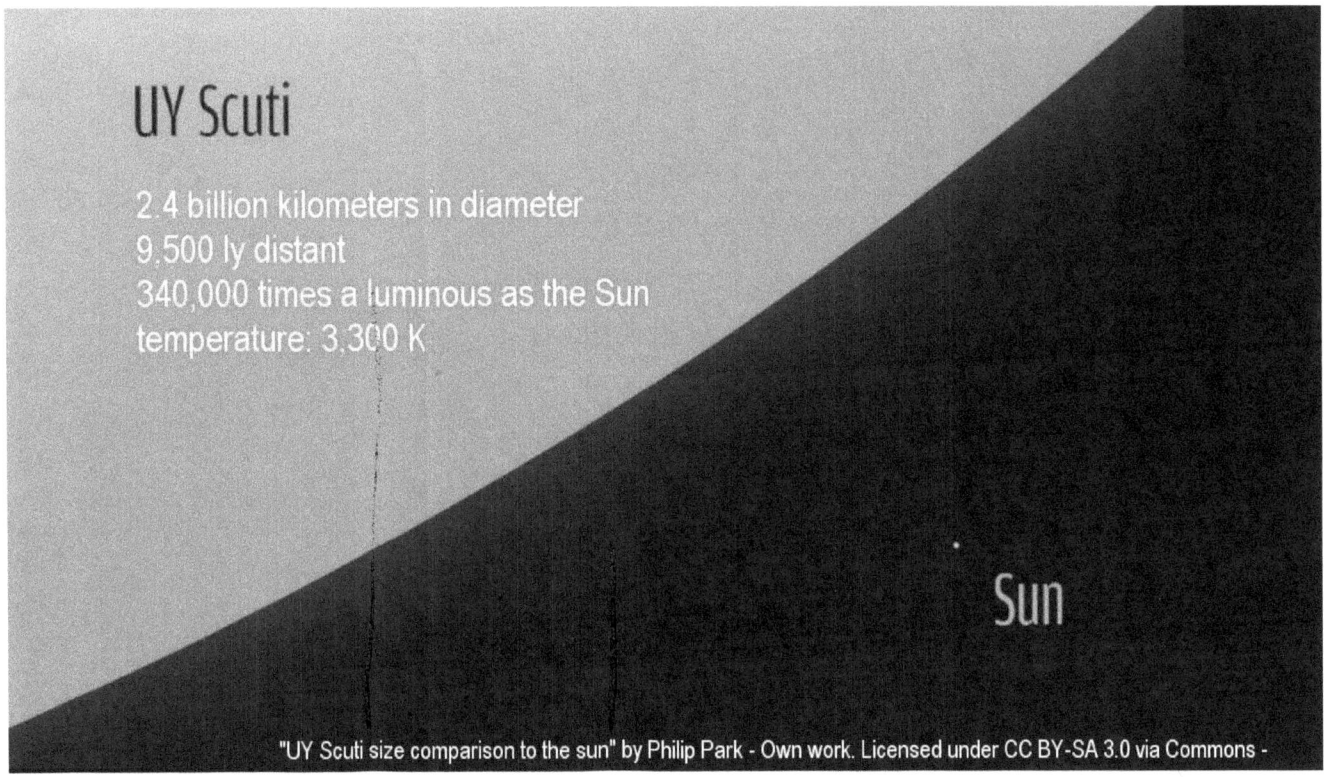

The paradox is further resolved if we consider the extremely large star UY Scuty, the largest star that is known. It is deemed to contain 10 times as much mass as our Sun. But this mass is dispersed in a sphere that is 1,700 times larger in diameter than our Sun, which makes it 5 billion times larger in volume.

In this 5 billion times larger volume, the 10 solar masses almost disappear. As one researcher has put it, the star's mass-density is so low, that it is practically a vacuum. This simple fact takes us far away from the gas-compression model of the Empire Sun, especially when we consider that this giant sphere of 10 solar masses, dispersed so thinly that it is almost a vacuum, is observed to be 340,000 times more luminous than our Sun.

Not possible under the gas-compression nuclear-fusion-sun model

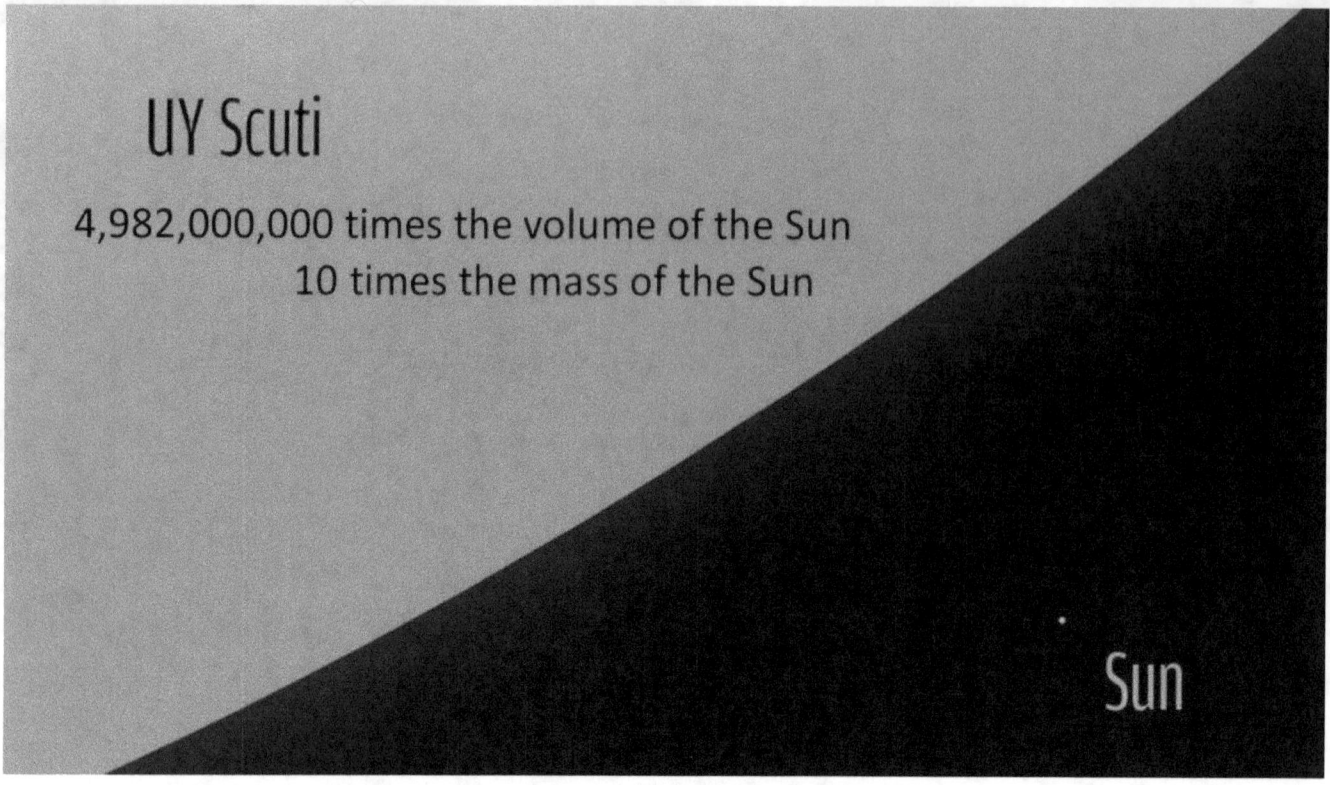

Such a giant star is not possible under the gas-compression nuclear-fusion-sun model. Paradoxically, it exists. We can see it. We can measure it. It is real. It exists, because as a plasma sun, the giant star is not paradoxical at all, but illustrates the natural dynamics of plasma physics.

It is totally possible under the plasma model for the largely-empty giant star of 10 solar masses, to outshine our Sun 340,000-fold,. This is so, because its radiated energy is not produced within it, but by plasma reactions occurring across its nearly 3 million times larger surface, than that of our Sun. That's where the radiated energy comes from. On this basis the numbers add up. The recognition enables us to break away from the Euclidian reductionist kind of thinking where everything is based on what one can see, where cause and effect are conceptionally isolated. Of course, the universe isn't isolated from itself. This myth exists only in the box of reductionist thinking. In real terms, the opposite is true.

➢ **The plasma 'sunshine' onto the Sun**

The plasma 'sunshine'

(onto the Sun)

The plasma 'sunshine' onto the Sun.

Just as streams of sunlight flood the earth, and illumine the landscape

Just as streams of sunlight flood the earth, and illumine the landscape, brighten the flowers, so stars are 'illumined' in cosmic space by plasma streams 'shining' unto them.

UY Scuti is immensely luminous as a huge canvas that has plasma 'shining' onto it

UY Scuti is immensely luminous, nine and a half-thousand light-years distant. It is that, because it exists as a huge canvas that has plasma 'shining' onto it, which it radiates back in the stellar 'colors' of atoms, light, and cosmic rays.

We need a university type of approach to be able to see with the mind

We need a university type of approach to be able to see with the mind past the reductionist mysticism where everything is deduced from the most primitive platforms and reduced to them, where nothing else is deemed to exist. That's like a person walking through the world with its eyes closed, who is thereby effectively blinded.

Inversely, with the eyes of the mind fully open, the real world comes into view with new vistas, vast opportunities, and revolutionary implications that otherwise remain unseen.

I chose the name Cygni for my own 'university' type explorations

The star NML Cygni
5,300 ly distant
1,183 times larger than the Sun
823 million km radius
1.65 billion times the volume of the Sun
20-50 times the mass of the Sun
Surface temperature: 3834 K
Luminosity: 272,000 times the Sun
Variable in 940 day intervals

Image by Nick Wright - Transferred from en.wikipedia to Commons by Hekerui., CC BY 3.0, https://commons.wikimedia.org/w/index.php?curid=22162669

I chose the name Cygni for my own 'university' type explorations beyond the fog of Euclidian reductionism, in reference to the star Cygni, the second-largest star known. It is 1100 times larger than our Sun and has up to 50 solar masses spread across its sphere that is 1.6 billion times larger in volume than the Sun, and outshines the Sun 272,000-fold.

However, Cygni is a star that no one has actually seen. It exists hidden in a dense nebula that is evidently of its own creating. Cygni can only be seen with the mind, and be detected with advanced instrumentation that expands the range of our perception.

The Empire-Sun poster has the opposite effect

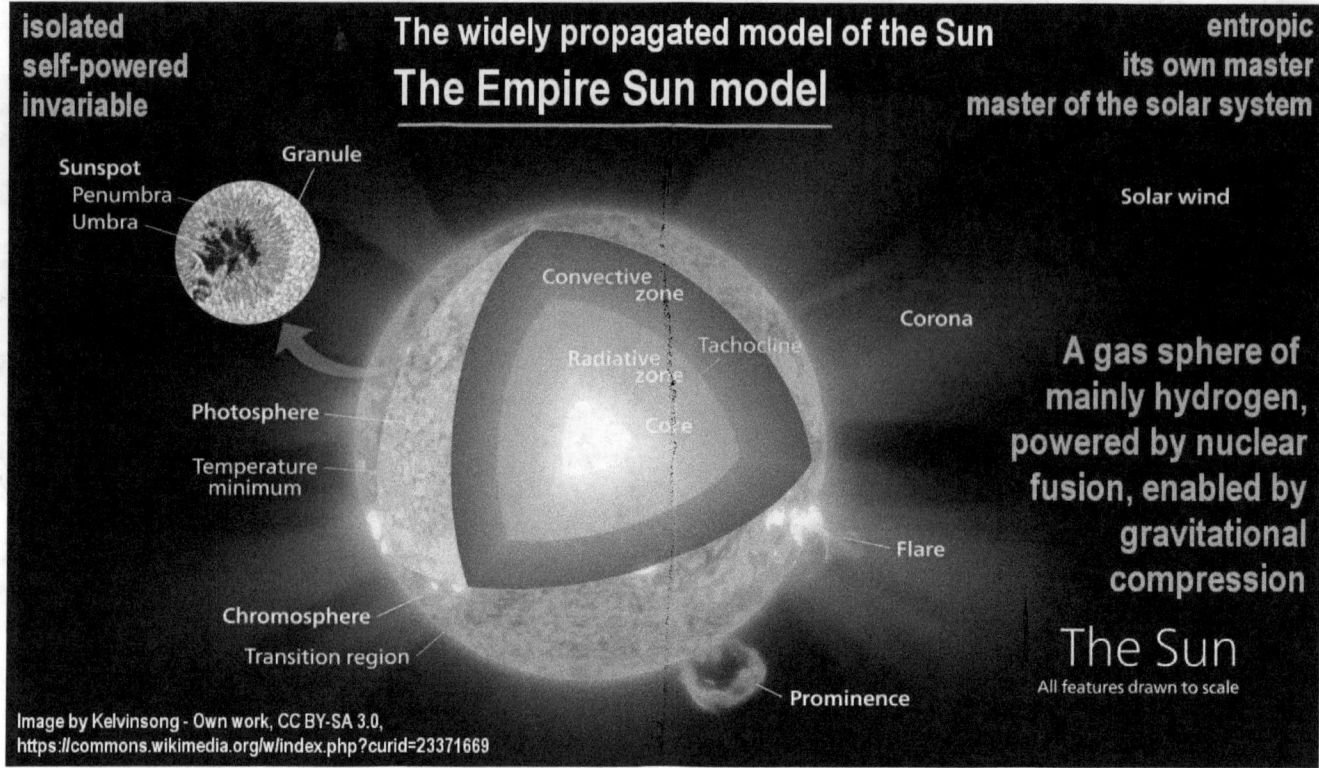

The Empire-Sun poster has the opposite effect. It takes the Sun out of the universe and then aims to explain how this this disconnected star is able to shine. The theory poses a problem and then solves the impossible with magic and mysticism where nothing is real.

Part 2: The Empire Sun: Its false face

➤ The Empire Sun: a trap

The Empire Sun

(a trap)

The Empire Sun: a trap

The effect of this trap is, that it reduces the cognitive power of humanity.

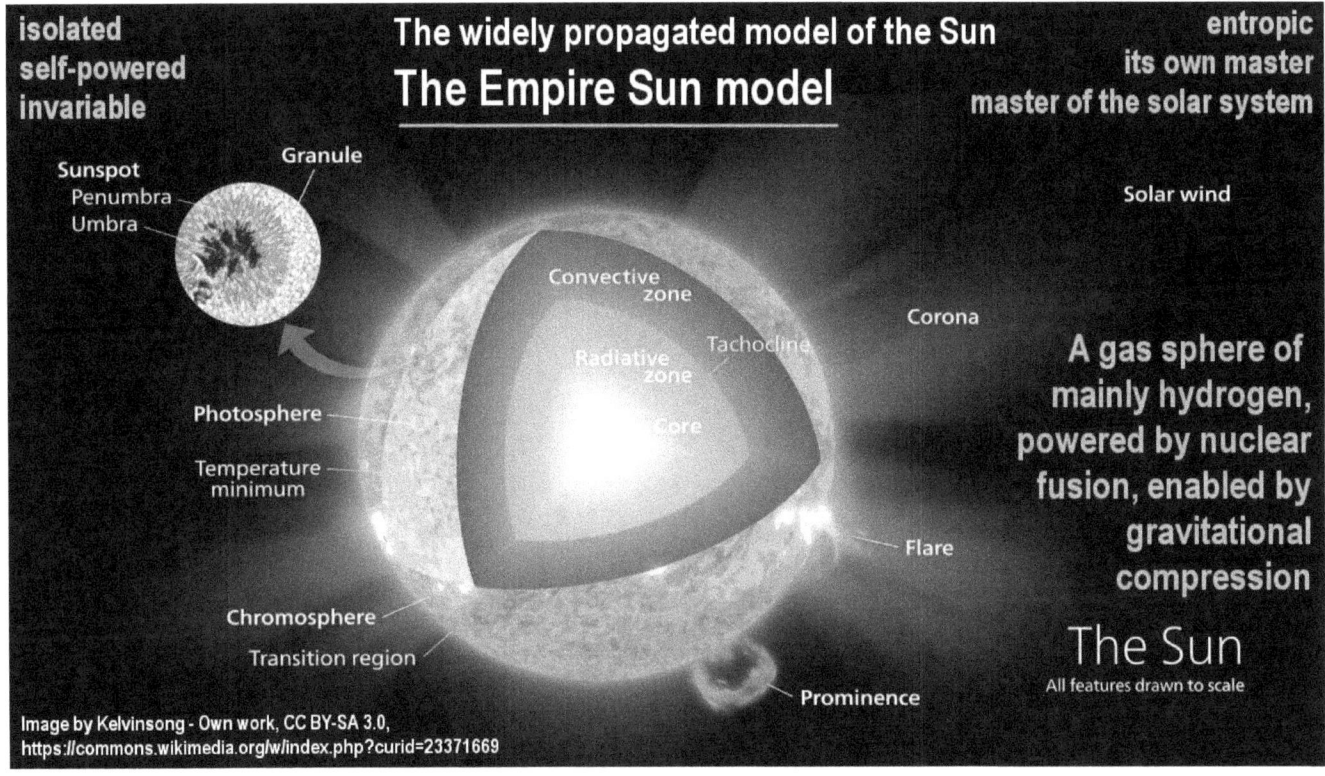

The effect of this trap is, that it reduces the cognitive power of humanity.

Reductionist theories, for which no evidence exists, nor is possible, revert the mind backwards to a more infantile state of thinking. The mind responds that way for its self-protection.

Mysticism that all structures of empire are built on, such as the later illuminaticism

Coat of arms of an emperor of the Holy Roman Empire

What we are confronted with here, is the kind of mysticism that all structures of empire are built on, such as the later illuminaticism and so on.

Empire creates conflicts in the mind for which no solutions are deemed possible

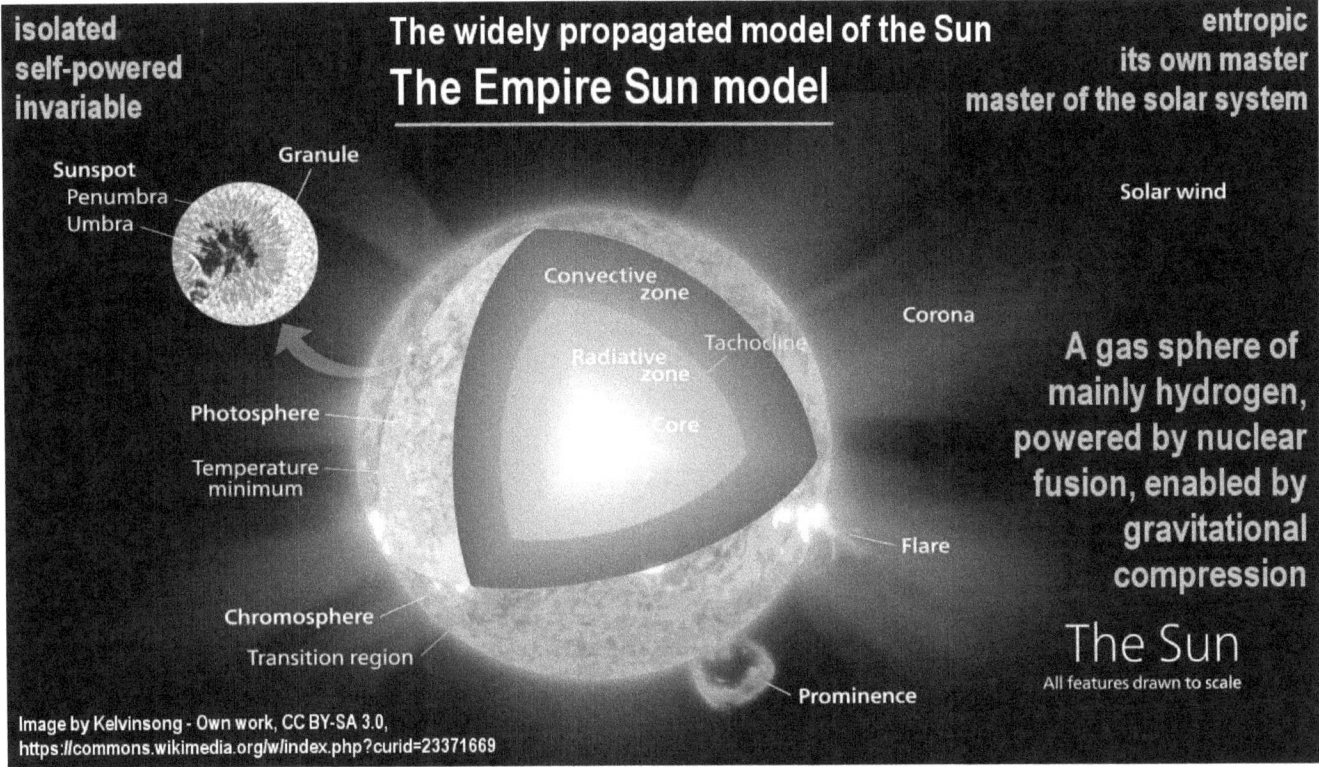

Empire creates conflicts in the mind for which no solutions, except magical solutions, are deemed possible. The result is that the reduction process renders the mind small and impotent. The impossible Sun that is conjured up on this basis, is an example of the mental emptiness that Empire imposes on society with which its defends its illegitimate existence.

On the political front, wars are achieved on the reductionist basis

Coat of arms of an emperor of the Holy Roman Empire

On the political front, wars are achieved on the reductionist basis. On the economic front, looting and depopulation are achieved by it. On the mental front, the war effort is focused on keeping the mind small and infantile, and thereby easily dominated.

Does this sound like a conspiracy? If so, it probably is. The system of Empire, which has no legitimate foundation to exist, lives by the forces of conspiracies of many types; the conspiracies for war, the conspiracies for stealing, the conspiracies to destroy civilization, the conspiracies for depopulation, and the conspiracies for fostering impotence.

➢ **Conspiracy for a false Sun, hiding reality, obscures that the Ice Age transition has begun**

Conspiracy for a false Sun, hiding reality

(obscuring that the Ice Age transition has begun)

Conspiracy for a false Sun, hiding reality, obscures that the Ice Age transition has begun.

The Empire Sun model is of the latter type type, that of the conspiracies

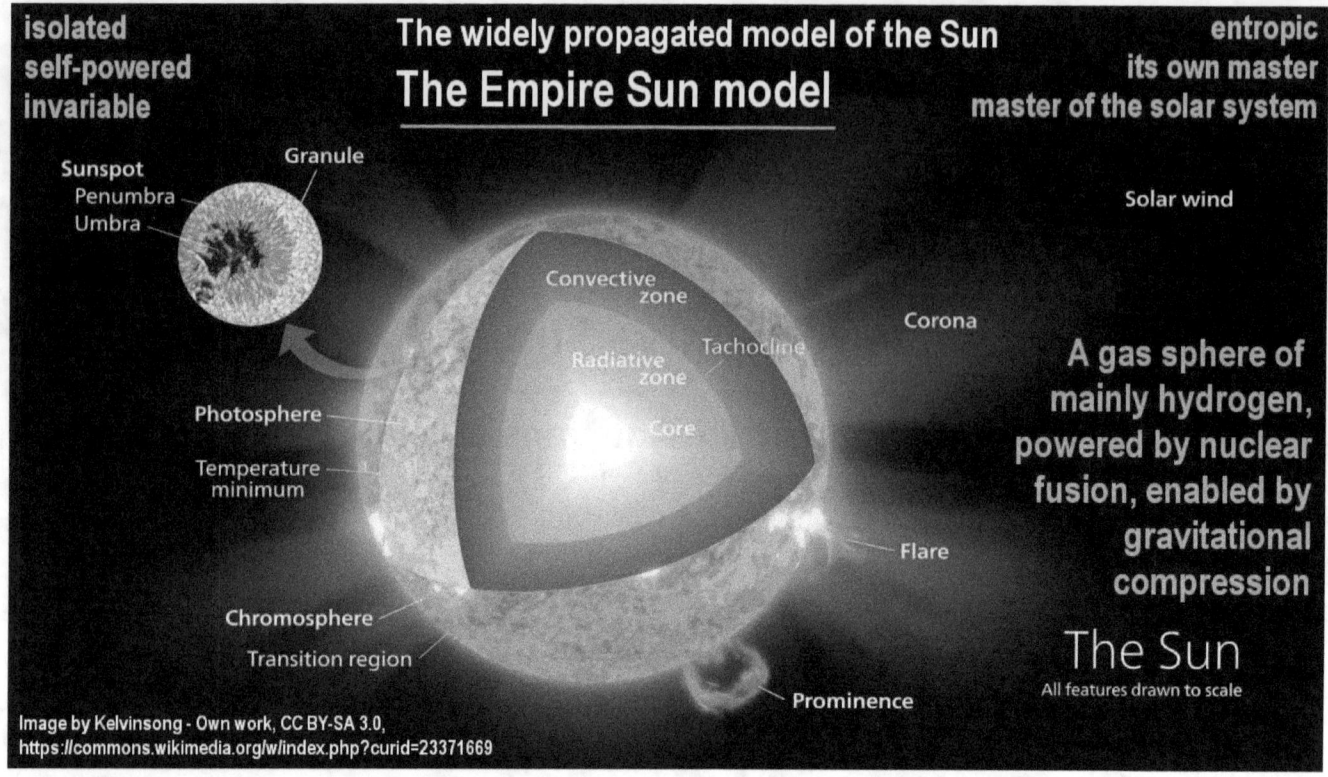

The Empire Sun model is of the latter type type, that of the conspiracies It is designed for generating docile impotence. However, the infantilism in thinking that it forges, is not a natural phenomenon of humanity. Consequently, society has the power to heal itself of this disease. Significant healing has indeed been achieved on many occasions in history when society roused itself to become more fully human.

On a much higher-level platform than the reductionist platform that Empire imposes

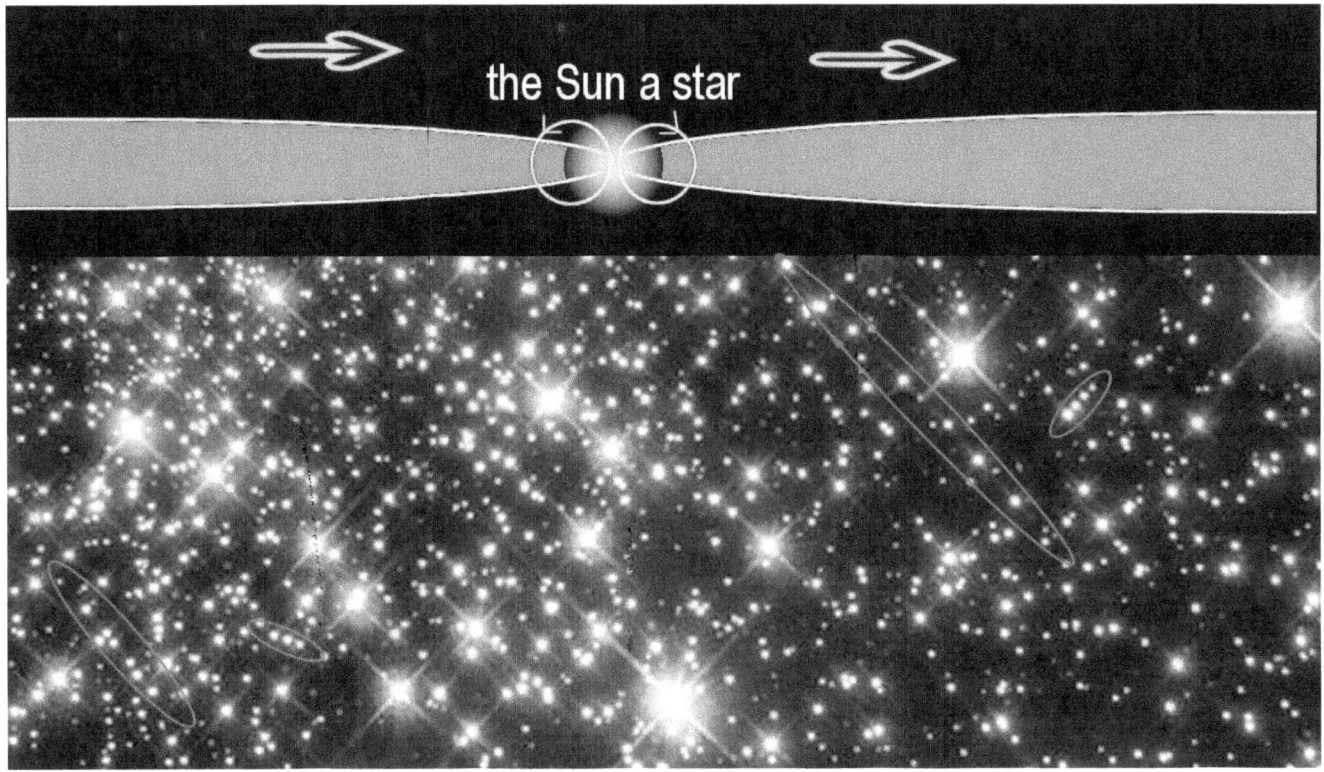

The real universe, of course, operates on a much higher-level platform than the reductionist platform that Empire imposes. On this higher-level platform everything is real and evidence for it exists everywhere.

In the cosmic universe every star is a plasma star. It interacts with with interstellar streams of plasma, which it attracts, which then become electromagnetically concentrated and focused onto the star by electromagnetic principles. When electrically charged particles are in motion in space, the electric movement creates a magnetic field around them. The magnetic force pinches the plasma stream into a smaller cross-section, which thereby increases the magnetic pinch effect, and so on.

The plasma pinching continues until the magnetic fields tangle up, flip backwards

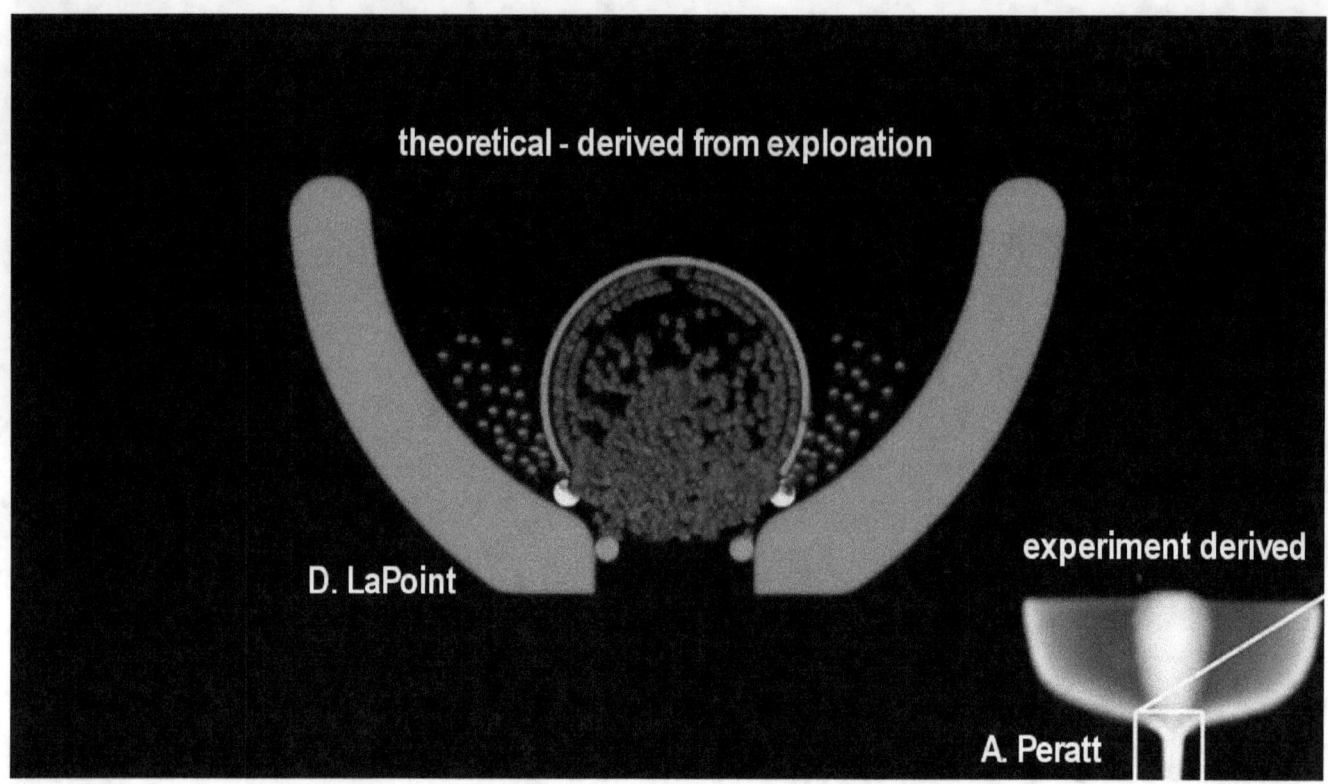

The plasma pinching continues until the magnetic fields tangle up, flip backwards, and thereby form a magnetic containment field under which the pinched plasma becomes densely concentrated.

The principle has been replicated in laboratory experiments, both statically, by David LaPoint, and dynamically, shown here in pink, produced at the Los Alamos National Laboratory in the USA under the direction of Anthony Peratt.

The concentrated plasma, under the confinement dome, becomes focused onto a sun

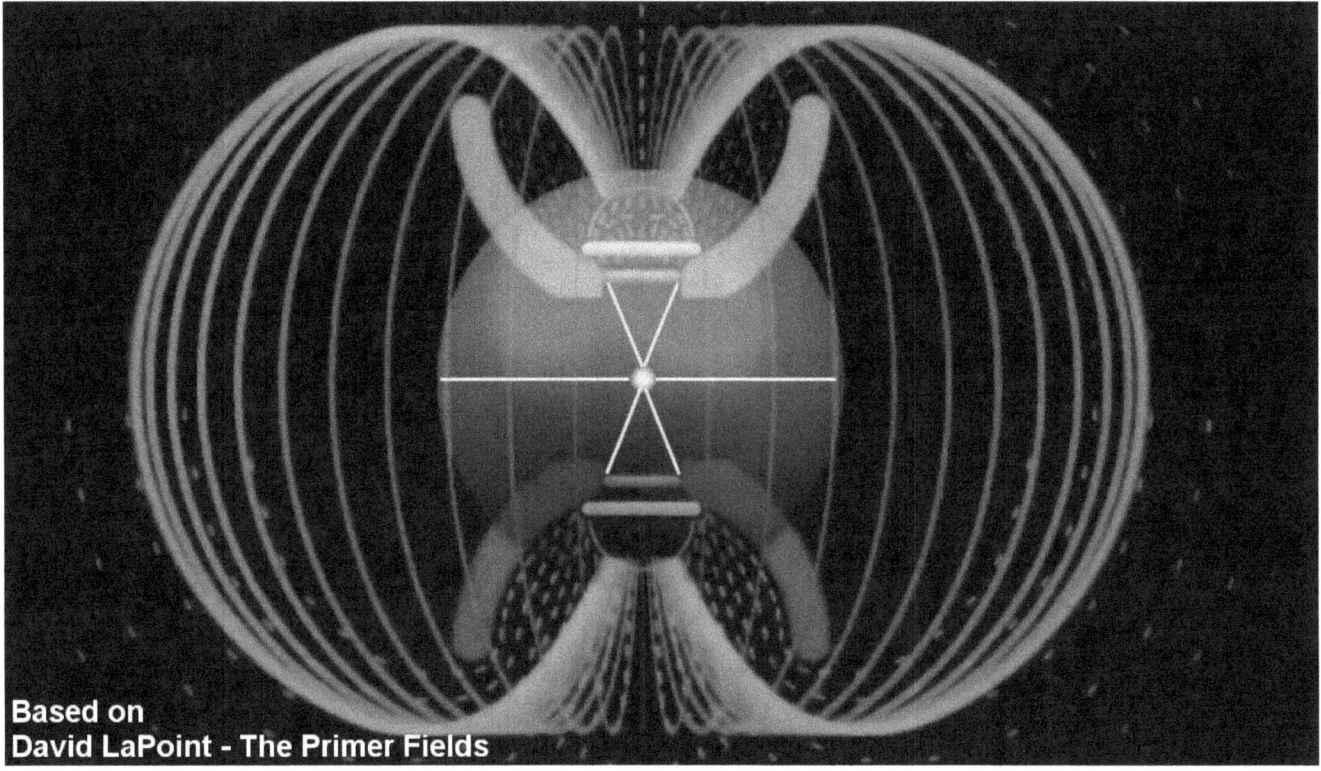

The concentrated plasma, under the confinement dome, becomes subsequently focused onto a sun. In a broad sense the magnetic-pinch principle and its plasma concentration illustrates the basic functioning of a solar system.

Interstellar plasma, flowing towards a sun; are concentrated magnetically. This in-flowing process is possible, because a part of the in-flowing plasma is consumed by the sun. The portion that is not consumed flows on to the next star by the inverse of the process that had concentrated it. David LaPoint termed these electromagnetic structures that facilitate this process, the Primer Fields.

We can see the complementary nature of the primer fields clearly evident

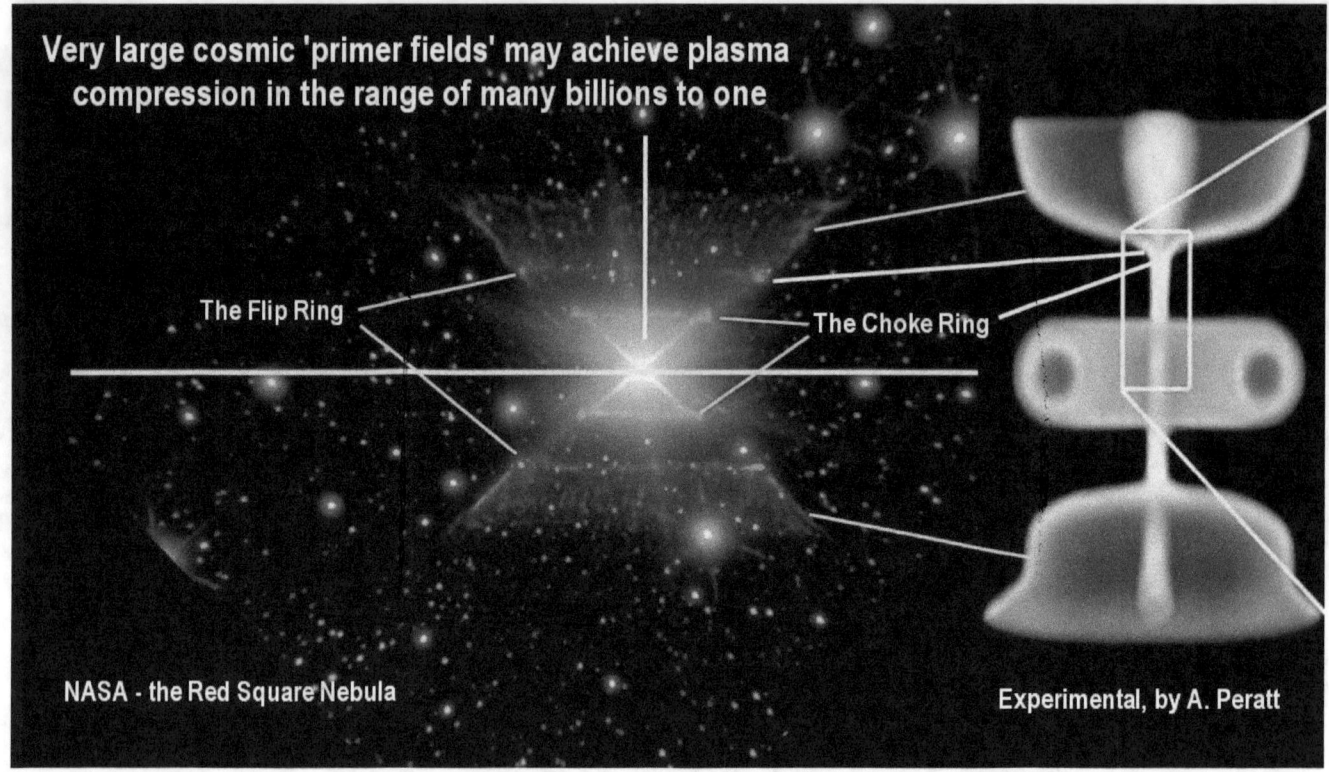

We can see the complementary nature of the primer fields clearly evident in space, as in the case of the Red Square Nebula shown here. We can also see it replicated in the dynamic laboratory experiment by Anthony Peratt.

That the Primer Field's principle applies to our Sun is evident in the measurements

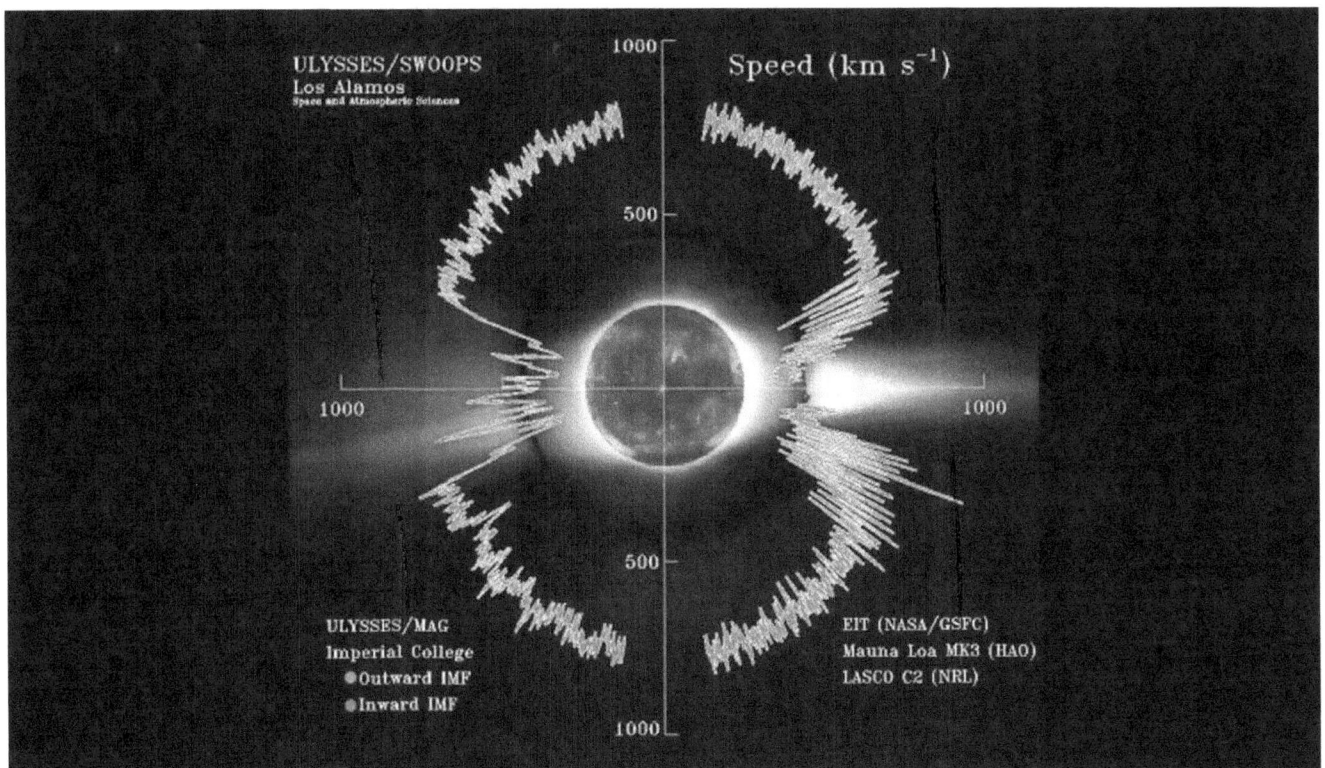

That the Primer Field's principle applies to our Sun is evident in the measurements by the Ulysses spacecraft that flew three orbits in a polar trajectory around the Sun. Whenever the spacecraft flew over the poles of the Sun, it encountered a void in the solar-wind pressure.

The voids were encountered exactly where the experimental models indicate

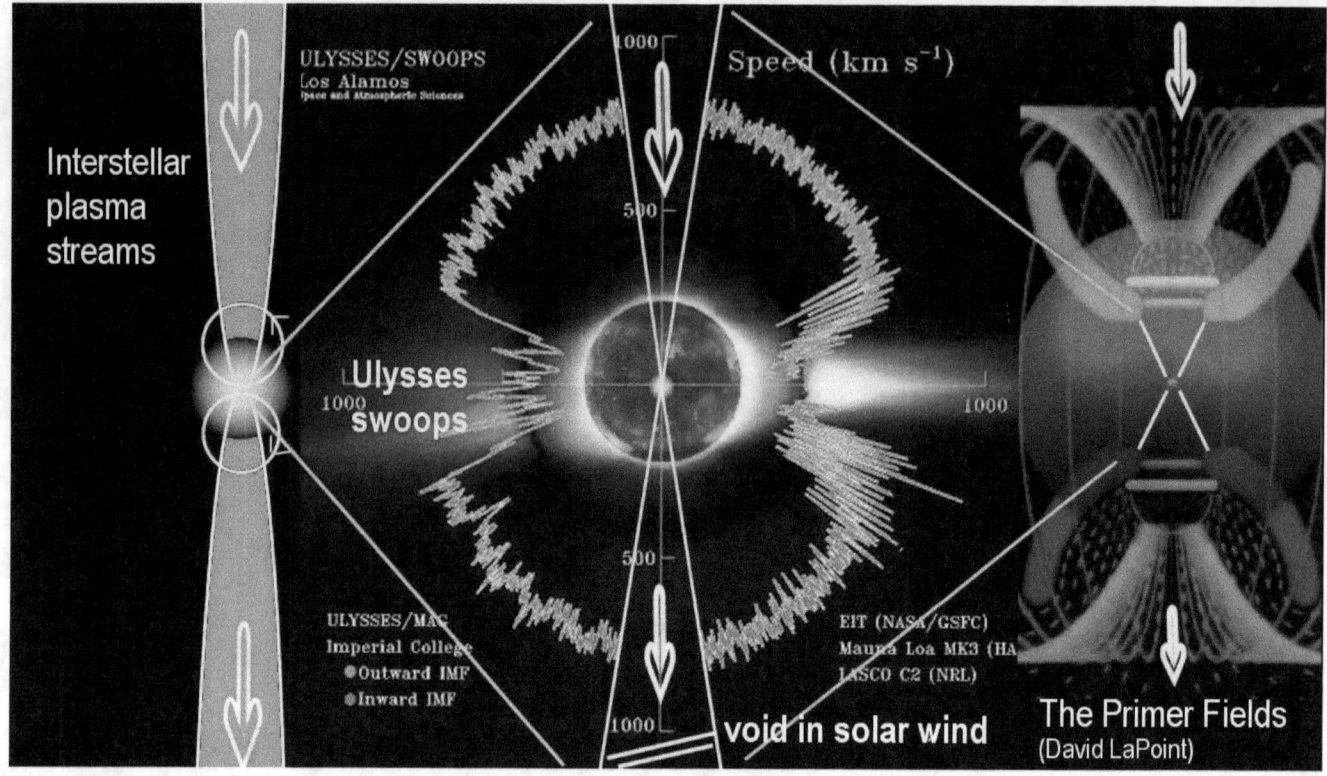

The voids were encountered exactly where the experimental models indicate that the plasma connection with the Primer Fields would be located.

The void that is encountered in space by Ulysses, accords with the result

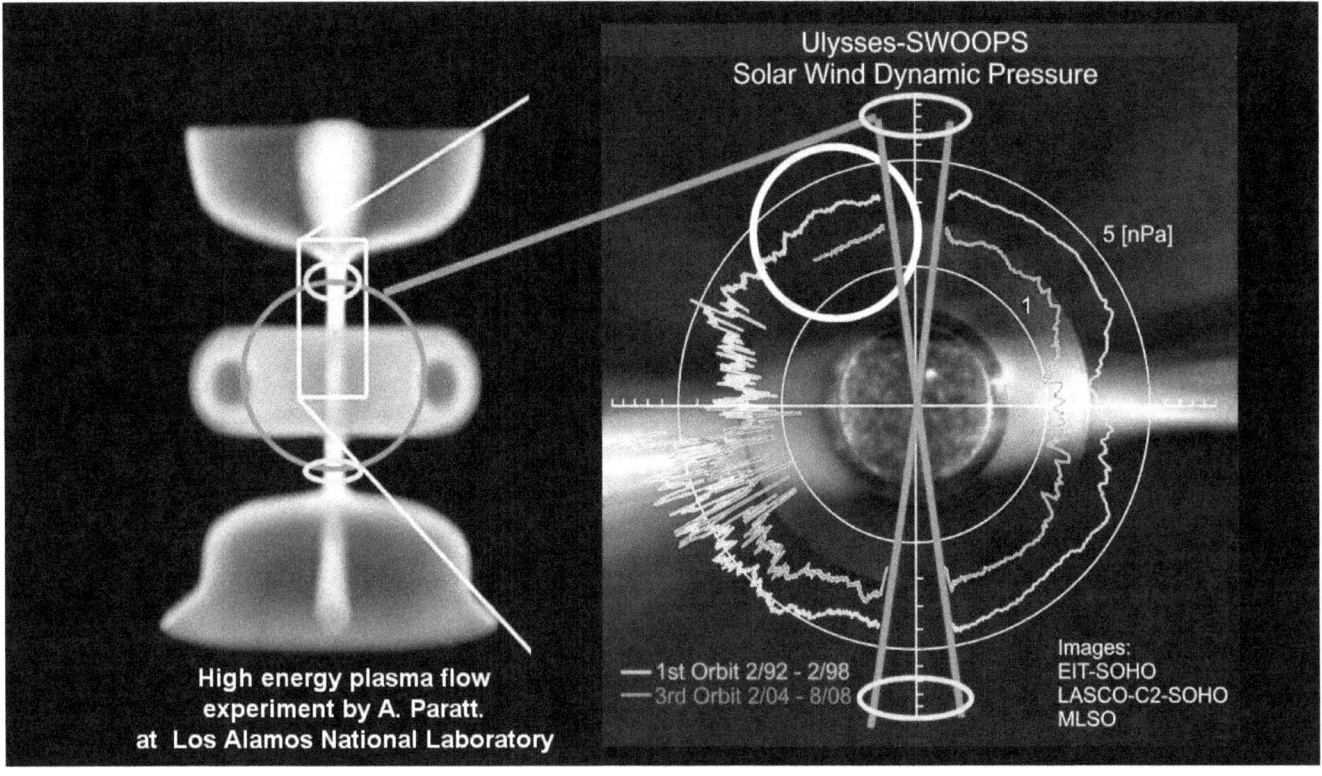

The void that is encountered in space by Ulysses, accords also with the result of the dynamic high-energy experiment by Anthony Peratt.

The effect that causes the Sun to be located at a node point

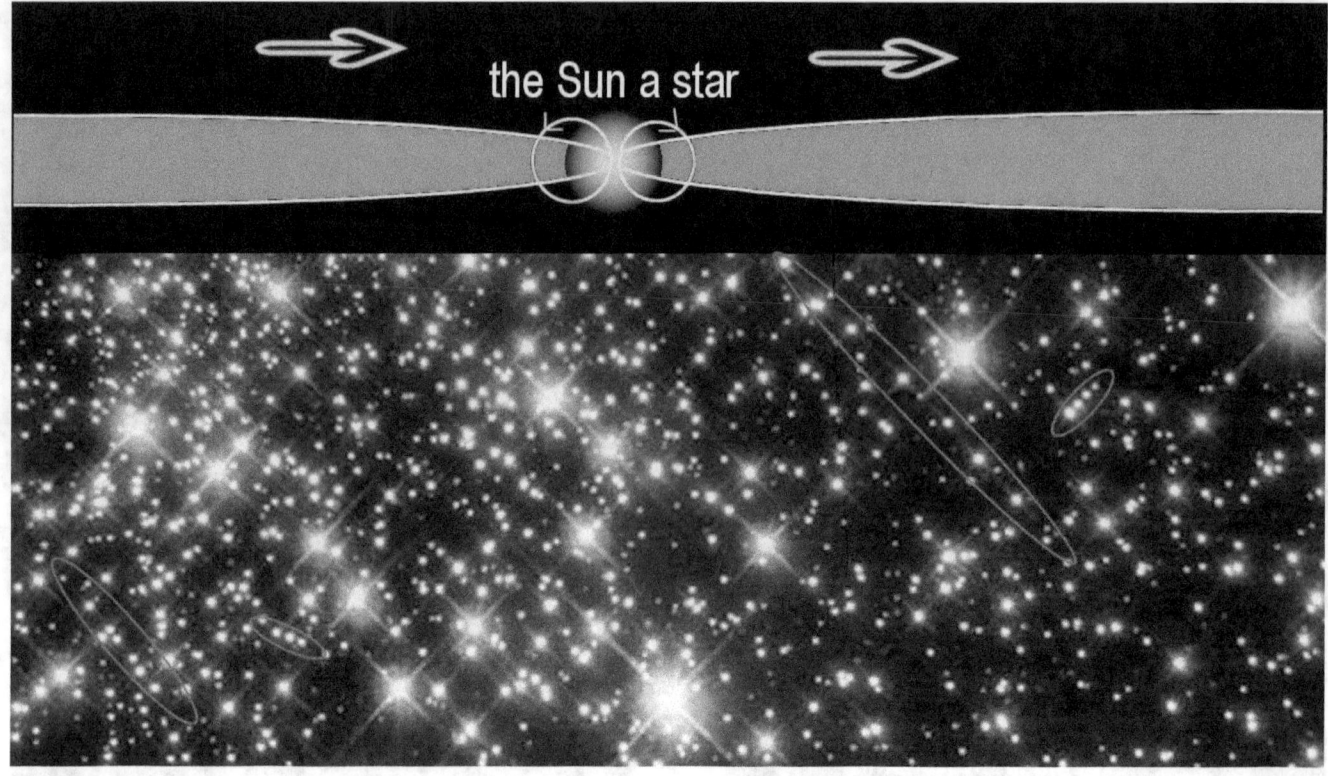

The effect that causes the Sun to be located at a node point in plasma streams is in part generated by the Sun itself. The cause is, that a sun, in principle, 'consumes' plasma.

In plasma, electrons and protons are free flowing

atoms are formed by the dynamic 'dance'
of electrons being attracted and forced to rebound

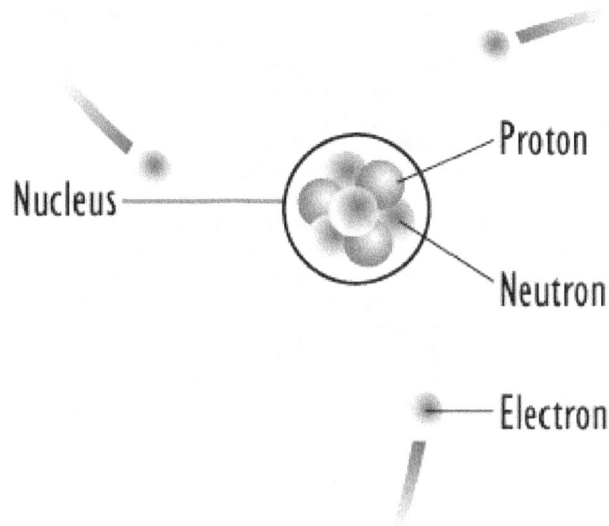

wikipedia (image)

In plasma, electrons and protons are free flowing. But in the surface layer of the Sun where the plasma concentration is extreme, plasma becomes dynamically bound together into atoms.

Atoms are electrically balanced structures of plasma that become electrically neutral

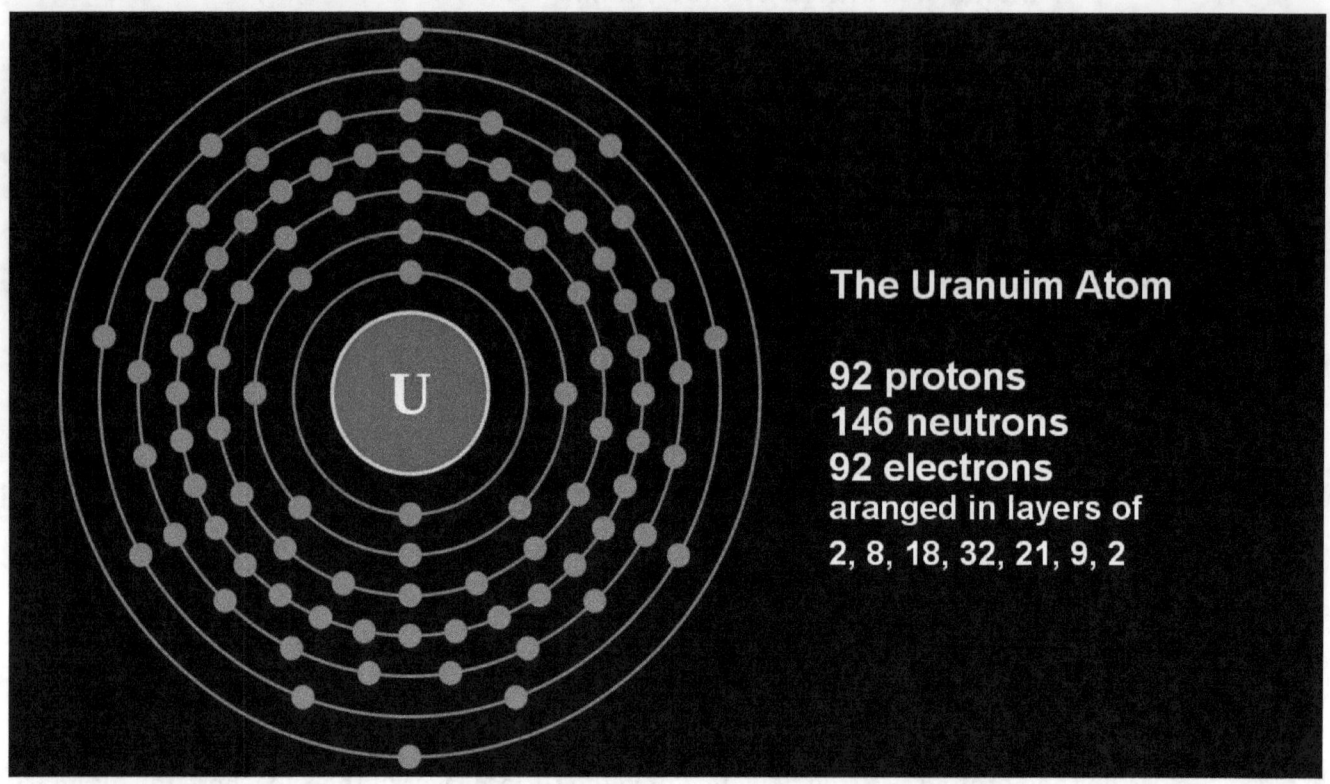

Atoms are electrically balanced structures of plasma that by the precise balancing become electrically neutral. By atoms being electrically neutral, the plasma that is bound up in forming the atoms, effectively vanishes from the electrodynamics landscape. All the atoms of the planets in the universe are created in this manner on the surface of a sun.

The synthesis of atoms creates the vital sink effects

The synthesis of atoms that creates electrically balanced packages, creates the vital sink effects that enables continuous streams of plasma to flow into a sun. The in-flow can only be maintained when there is an 'outflow' happening into what is termed, a sink.

The synthesizing process furnishes the sink effect. The synthesized atoms, by their electric neutrality, are no longer a part of the electric plasma system. The plasma particles that created the atoms no longer exist in unbound form. They are thereby free to simply flow away with the wind, which in this case, is the solar wind.

➢ Sunlight NOT from the Sun

Sunlight

(NOT from the Sun)

Sunlight NOT from the Sun

The atoms that are forged on the surface of the Sun

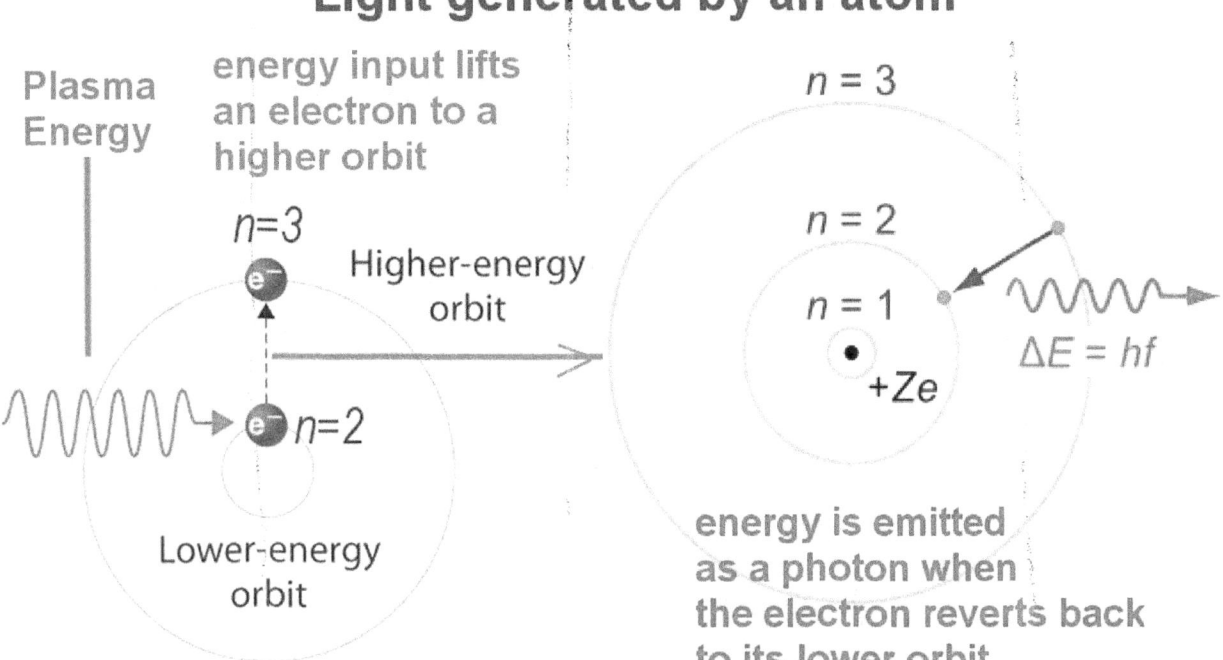

The atoms that are forged on the surface of the Sun. They serve as a catalyst that converts plasma energy into photons of light and heat that every sun or star radiates.

The energy the we receive on Earth from the Sun as light and heat, is plasma energy converted into photons by atomic elements that are synthesized on the surface of the Sun.

It takes a wide-ranging assortment of different atomic elements

It takes a wide-ranging assortment of different atomic elements, all acting in their own way, generating light at their own unique wavelengths, for the white symphony of light to flow together that contains the seamless band of colors that we see as sunlight.

This 'perfect' balance isn't possible under the Empire Sun model

This 'perfect' balance isn't possible under the Empire Sun model that renders the Sun a hydrogen gas star, but is possible to be produced by a plasma sun that contains all the natural atomic elements on its surface.

Of course, the Plasma Sun emits not only atoms and light

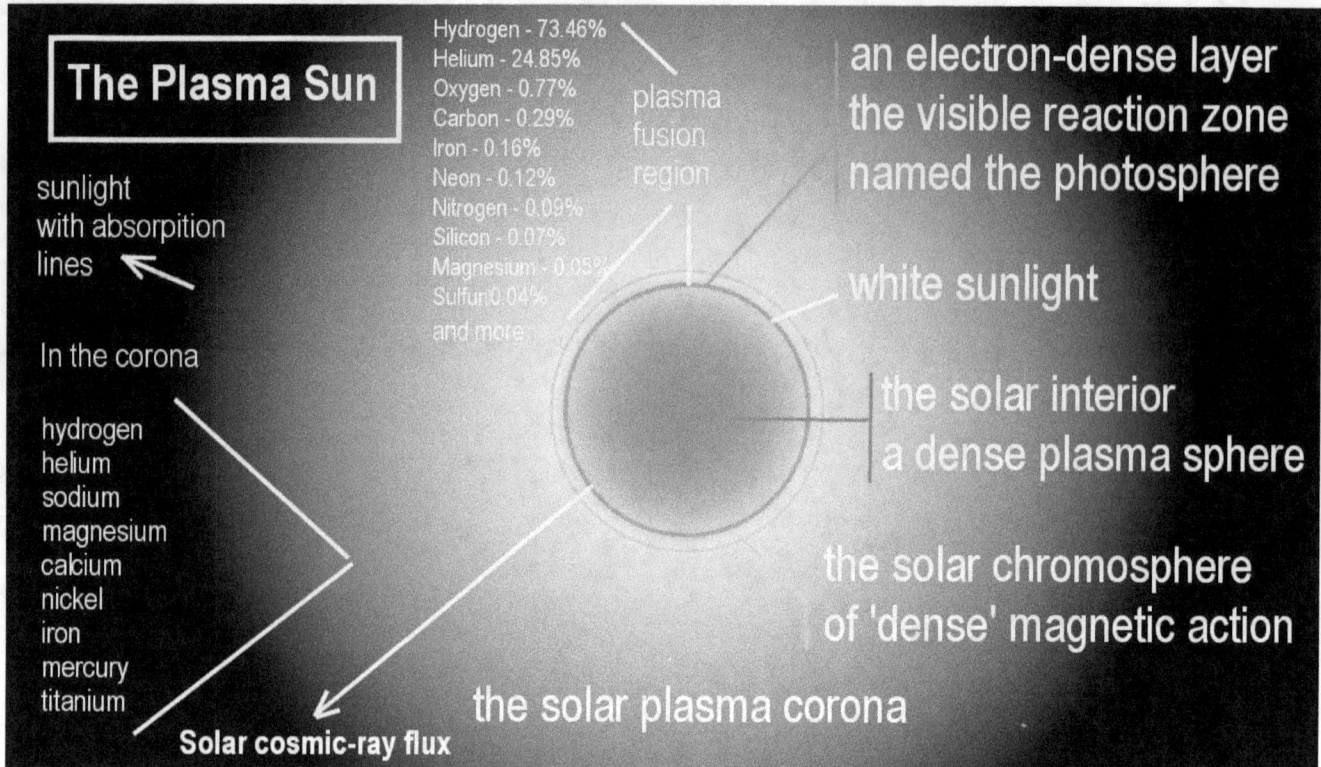

Of course, the Plasma Sun emits not only atoms and light. It also emits solar wind and solar cosmic rays, neither of which are possible either, under the Empire Sun model that has its atomic reactions occurring deep within it under a mantel of primarily hydrogen gas half a million kilometers thick, as the theory stipulates. But with the Plasma Sun, that has its atomic reactions occurring at its surface, the emission of solar wind and cosmic rays is inherently natural.

The solar wind is comparable to a kettle letting off steam

Maximum temperature of liquid water at ambient pressure is 100 degrees Celsius: The Boiling Point

The solar wind is comparable to a kettle letting off steam. The more energy flows into it, the more steam flows out of it. On the Sun the kettle is the fusion cell.

The surface of the Sun is a vast sea of 'granular' fusion cells

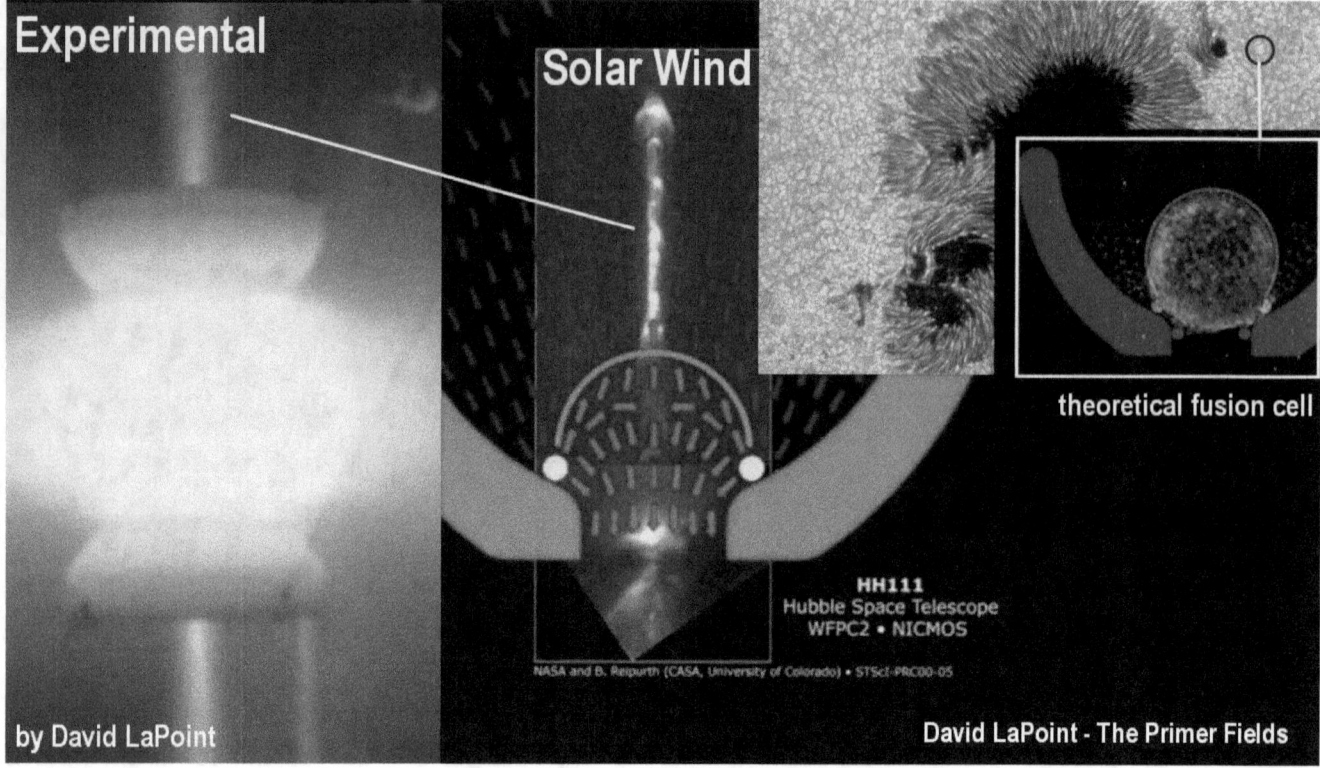

The surface of the Sun is a vast sea of 'granular' fusion cells, each with its own primer field structure.

Each surface primer filed has its own magnetic confinement dome, as one would expect, that enables the final concentration of plasma for the fusing of atomic elements to become possible. However, when the plasma pressure in the corona of the Sun enables a greater rate of flow than the fusion process can 'consume,' the excess is vented at the weakest point of the confinement dome in a jet of plasma that becomes the solar wind.

The excess pressure is released in concentrated form

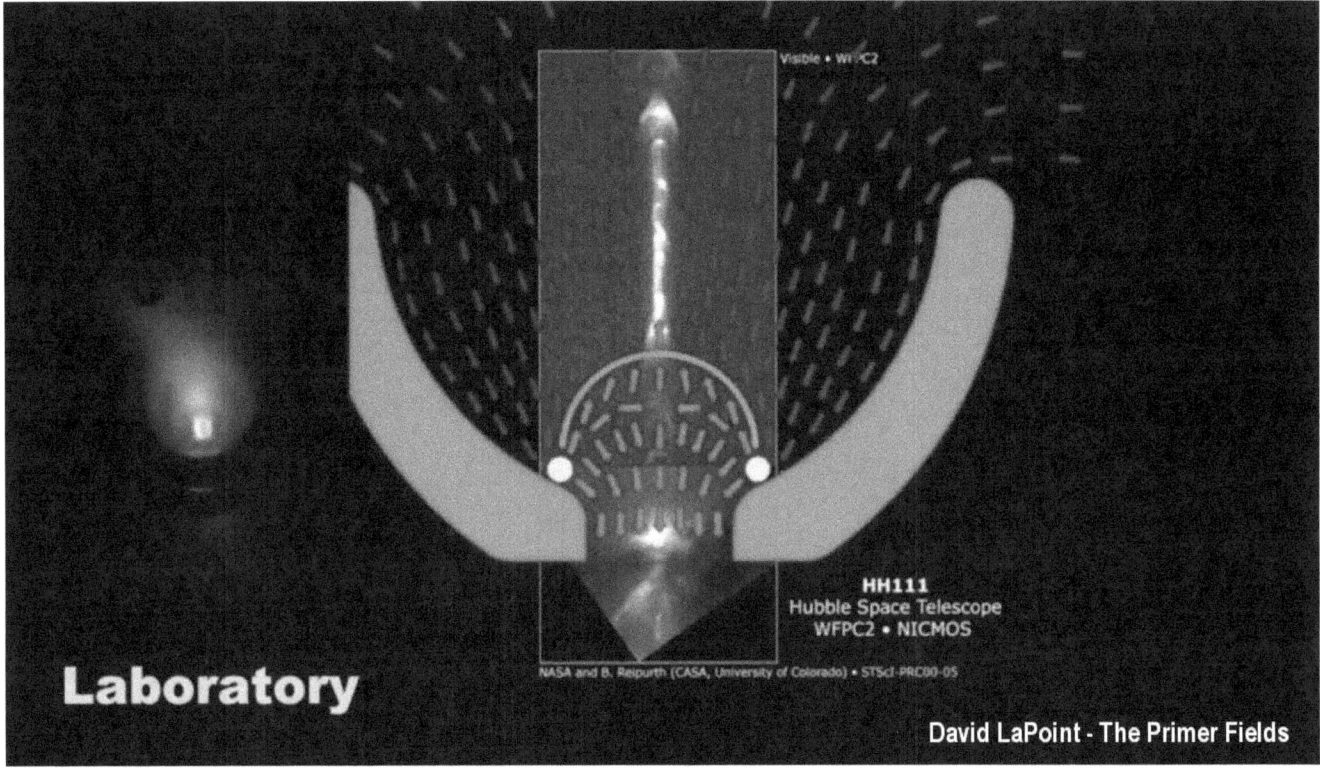

The excess pressure is released in concentrated form. The released plasma expands by electric expansion and thereby accelerates itself away from the Sun by the electric force in a similar manner as a bullet is propelled in a gun by the expanding fire of gun powder.

The expanding plasma becomes the solar wind that flows away from the Sun

http://www.zam.fme.vutbr.cz/~druck/Eclipse/ - an example of the amazing solar eclipse photography of Milloslav Druckmueller

The expanding plasma becomes the solar wind that flows away from the Sun. The strength of the solar wind gives us a measurable indicator of the density of the plasma surrounding the Sun.

Plasma is invisible. It does not emit light. However, the solar wind is measurable, even a long distance from the Sun. The stronger the wind flows, the stronger is the plasma sphere surrounding the Sun that drives the process.

The solar wind is able to flow out through the solar corona

The solar wind is able to flow out through the solar corona, which is inflowing towards the Sun, because in electrodynamics, electric currents flowing in opposite directions repel each other. The solar corona thereby becomes a 2-way street.

The solar fusion cells also emit cosmic rays

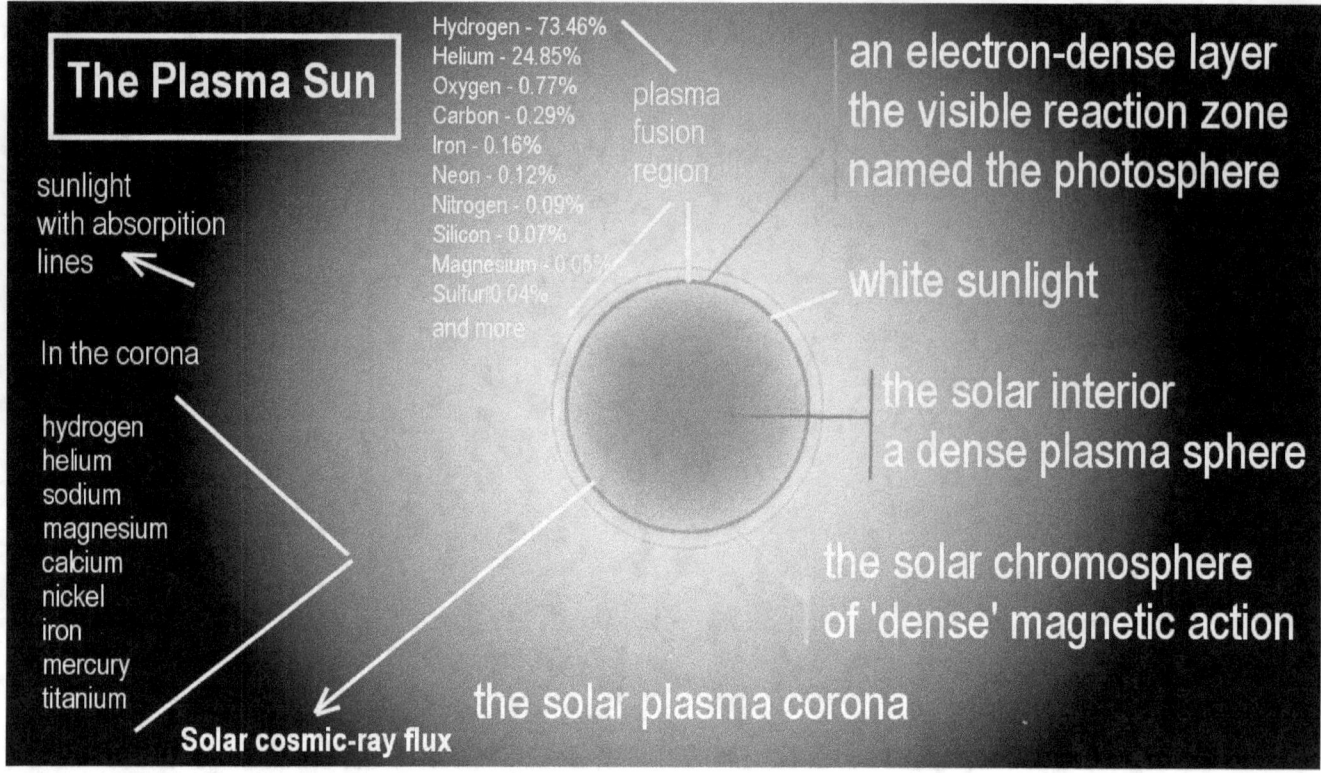

The solar fusion cells also emit cosmic rays, which are extremely energized protons and electrons that exceed the requirements for atomic synthesis. When they escape the fusion cells, they escape as fast moving cosmic objects, termed cosmic rays.

While most of the escaping cosmic rays get trapped in the solar corona

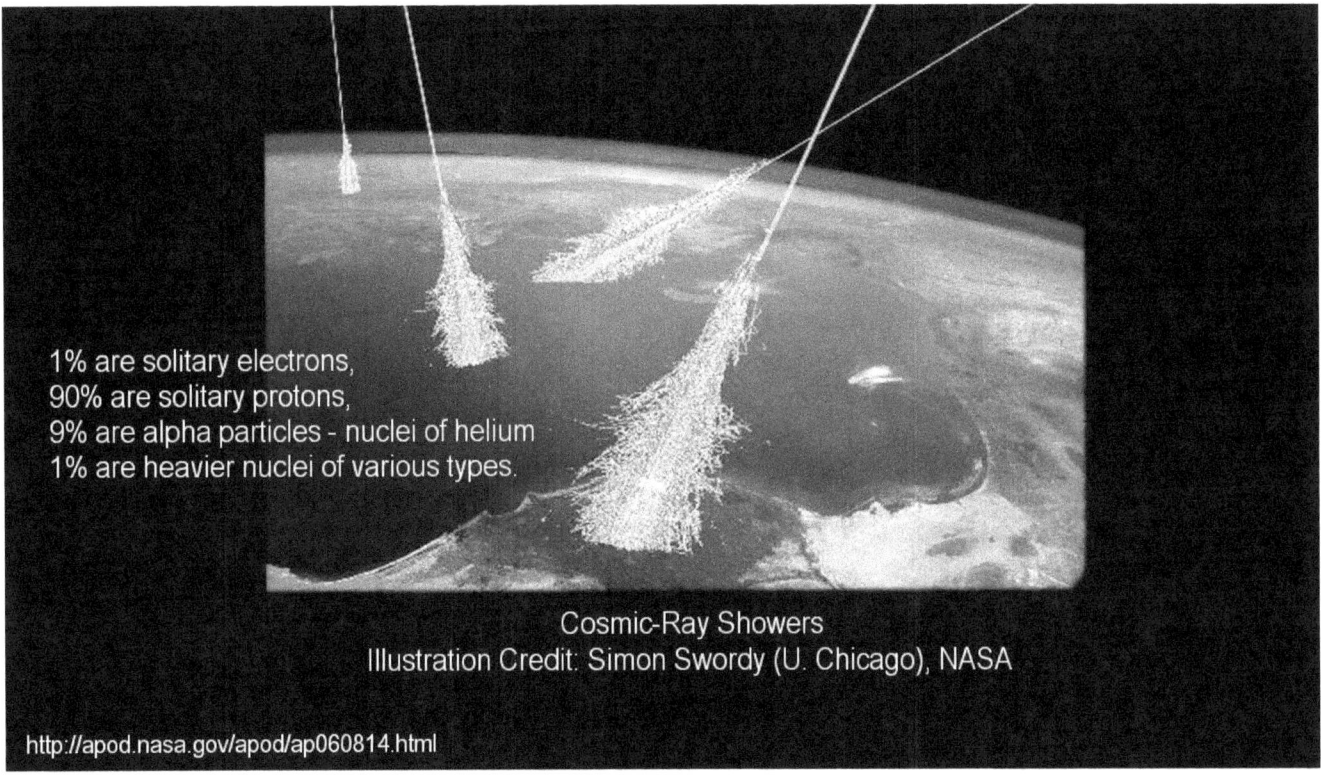

While most of the escaping cosmic rays - from the fusion cells - get trapped in the solar corona, a few get through the barrier and shower the Earth with wide-ranging effects.

Some of the effects are more amazing than we care to imagine.

The science researcher Simon Snoll had conducted a series of reaction-experiments

The science researcher Simon Snoll had conducted a series of chemical and biological reaction-experiments that were repeated continuously for days in an identical manner. One would assume that the measured rate of reactivity would be identical too, in every case, for the identical experiments. That's not what happened.

New Paradigm for Mankind - Cosmophysical Factors in the small

https://www.youtube.com/watch?v=wkDY_8HjMfk

The measured results always varied

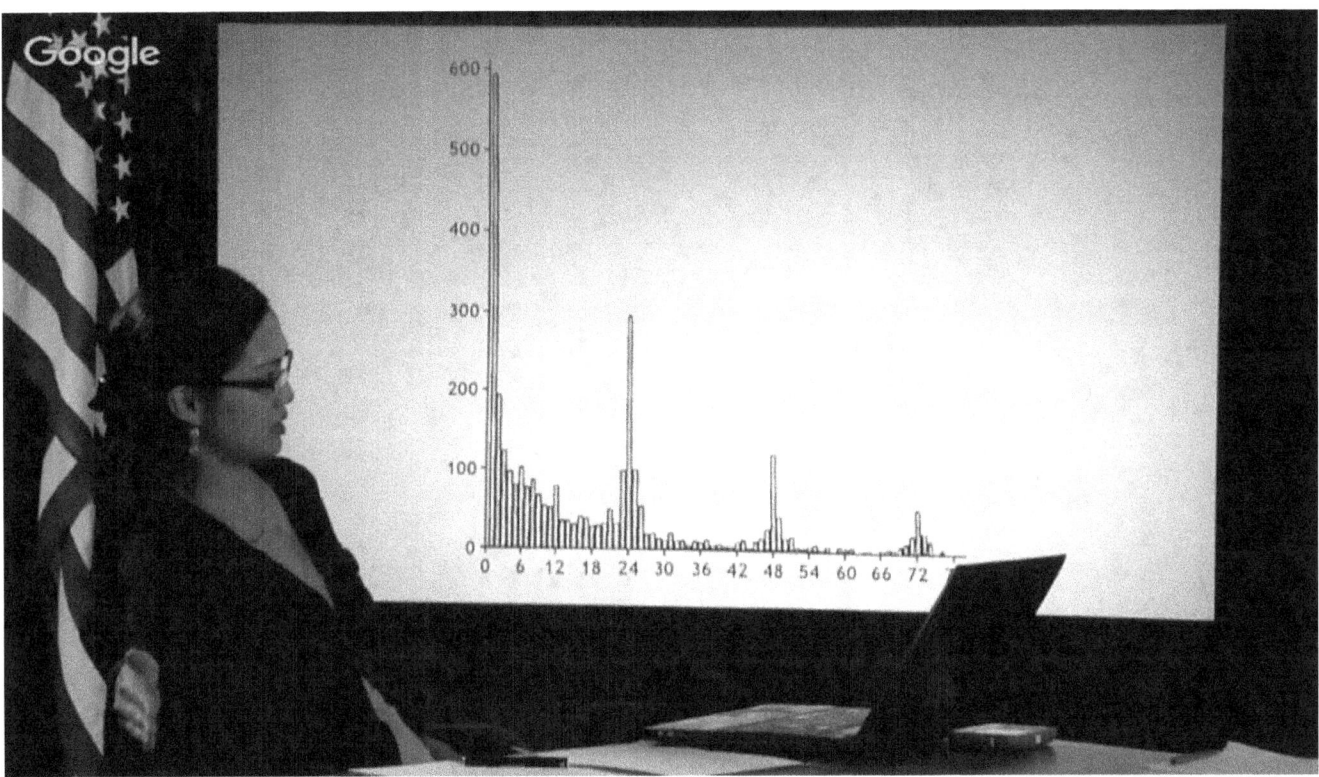

The measured results always varied. In one long-extended series of experiments the reaction rate had spiked every 24 hours, with the spikes themselves becoming smaller.

The result makes no sense, except when one notes that the spikes occur with the periodicity of the rotation of the Earth and that the successive spikes became smaller.

This is the pattern that one would expect to see when a coronal hole opens up on the Sun in the direction of the Earth through which larger volumes of cosmic rays escape and shower our planet. The volume of the escaping cosmic rays received on Earth would naturally diminish over the span of four days as the coronal hole slowly rotates away from the Earth with the rotation of the Sun.

The experiment proves positive that solar cosmic rays are a real, are a measurable feature, and have a larger effect on Earth than one might expect.

New Paradigm for Mankind - Cosmophysical Factors in the small

https://www.youtube.com/watch?v=wkDY_8HjMfk

Cosmic rays also affect our atmosphere, in almost the same manner

ISS-34 - Stratocumulus clouds

Cosmic rays also affect our atmosphere, in almost the same manner. When cosmic rays flow through the Earth's atmosphere their electric quality ionizes atoms on the path of their encounter, which makes them up to 100-times more attractive to water vapor. The increased attraction increases the rate of cloud nucleation. Increased cloudiness makes the Earth colder, because the white surface of clouds reflects a portion of the incoming solar energy back into space.

By this process, the weakening of the solar corona causes the climate on Earth to become colder, because the weaker corona allows more cosmic rays to escape from the Sun, which increases cloudiness on Earth.

While the increased cloudiness that we now see evermore of, is NOT ultimately causing an Ice Age to occur, which has a deeper cause, the increasing cloudiness is nevertheless a significant indicator that the Sun is getting weaker.

Another indicator that the Sun is getting weaker

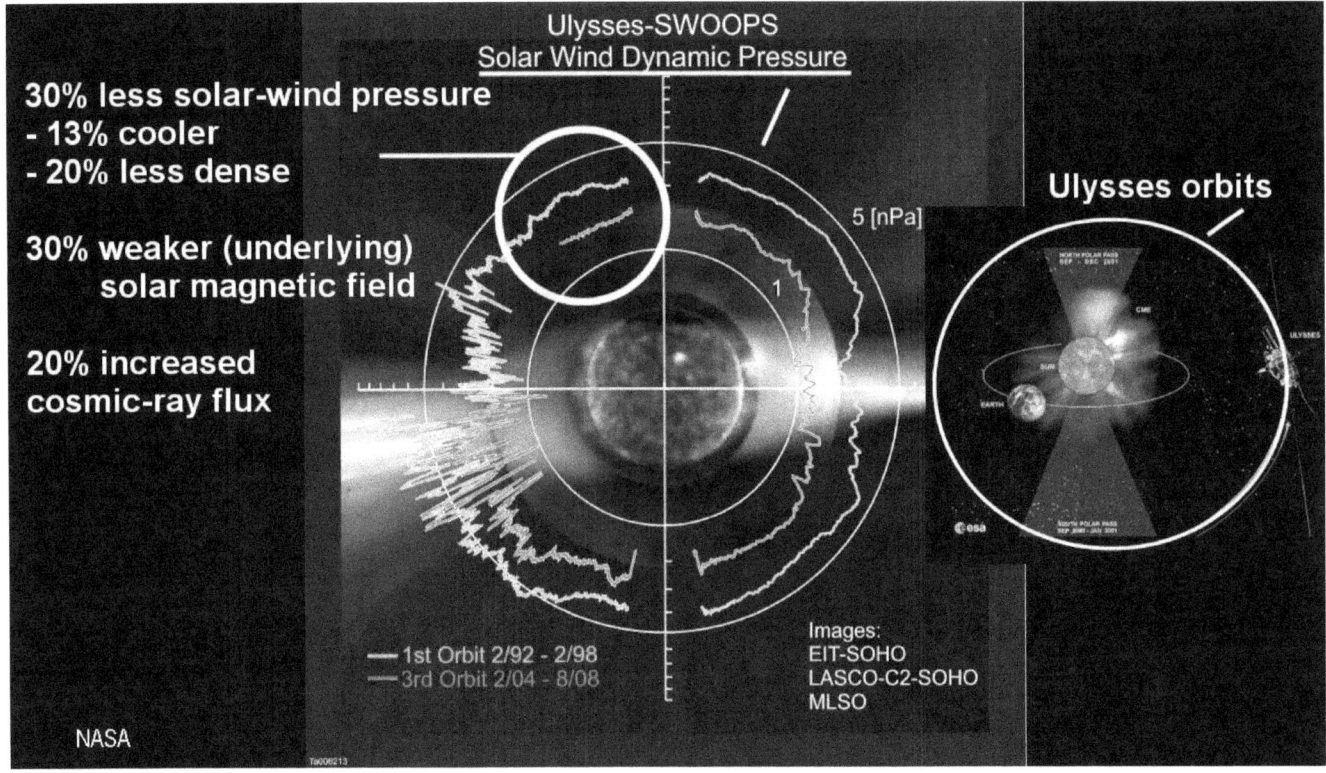

Another indicator that the Sun is getting weaker is the rapidly diminishing solar-wind pressure that the Ulysses satellite has measured the beginning of. The Ulysses satellite thereby measured the status of the solar system as a whole and its ongoing diminishing energy density.

➢ What masters the Sun?

What masters the Sun?

What masters the Sun?

Numerous forms of evidence prove that the Sun is not its own master

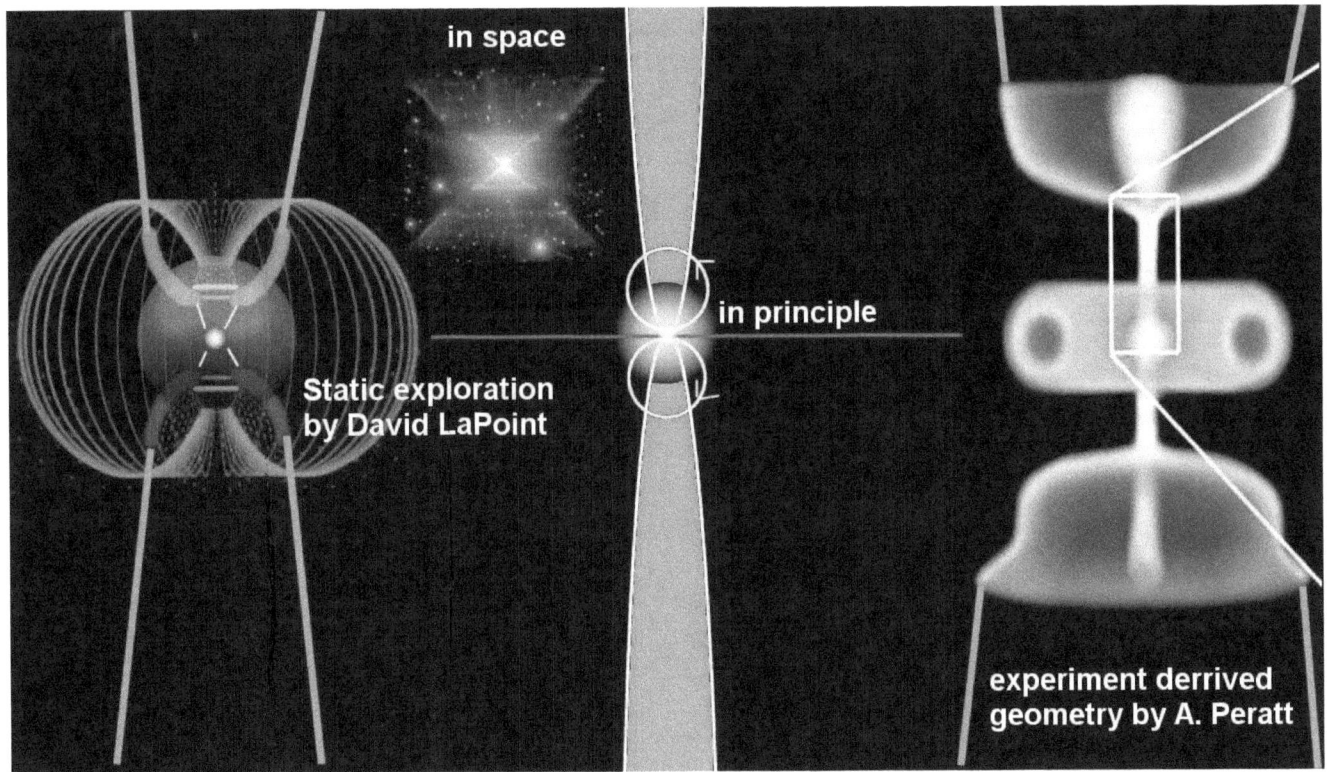

The numerous forms of evidence prove that the Sun is not its own master, but merely reflects, as a catalyst, the density of interstellar plasma streams. The plasma density that flows into the solar system from these very long interstellar plasma streams is in turn affected by electric resonance effects in the plasma streams, and is also affected by the general condition of the galaxy as a whole.

All of these variable factors, which affects our Sun from areas deep in space

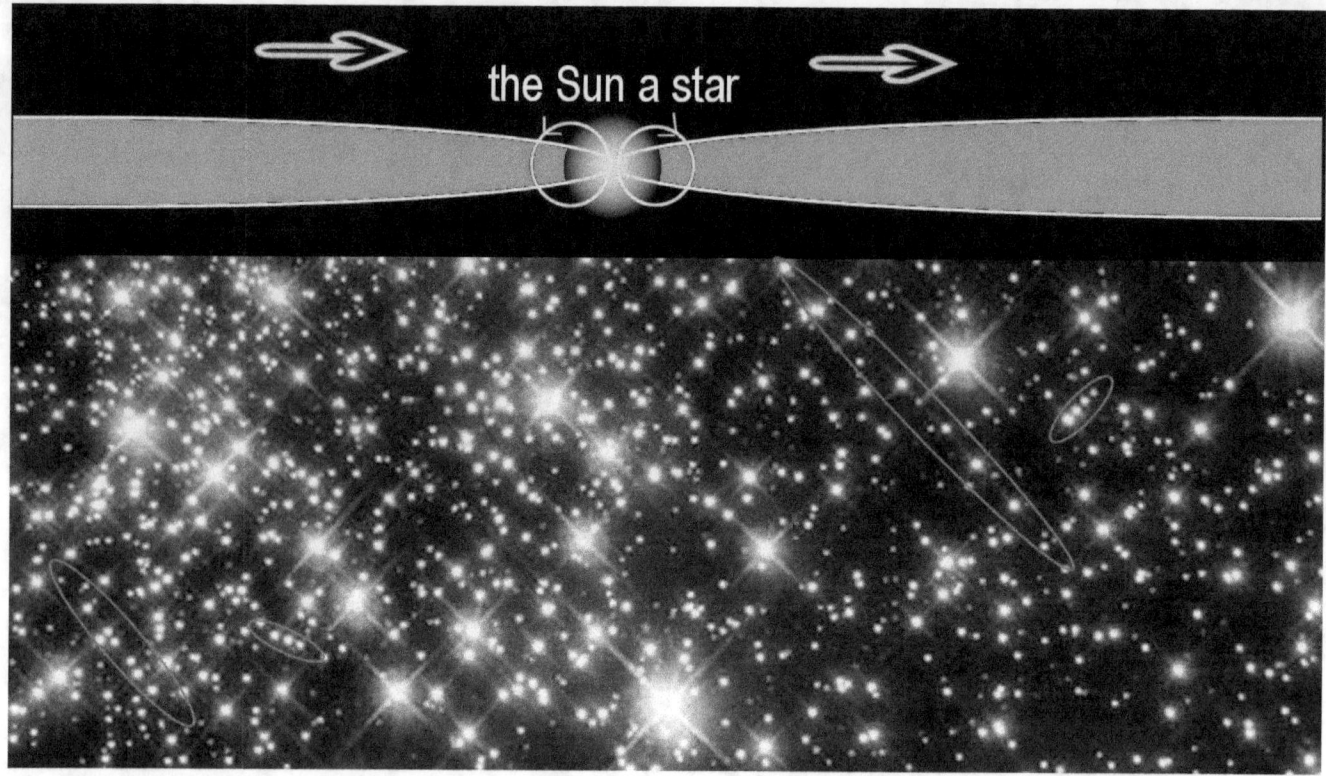

All of these variable factors, which affects our Sun from areas deep in space, secondarily affect the climate on Earth. Our climate is thereby imposed on us from afar by the numerous factors that affect our Sun.

Our Galaxy is presently at its weakest state in 440 million years

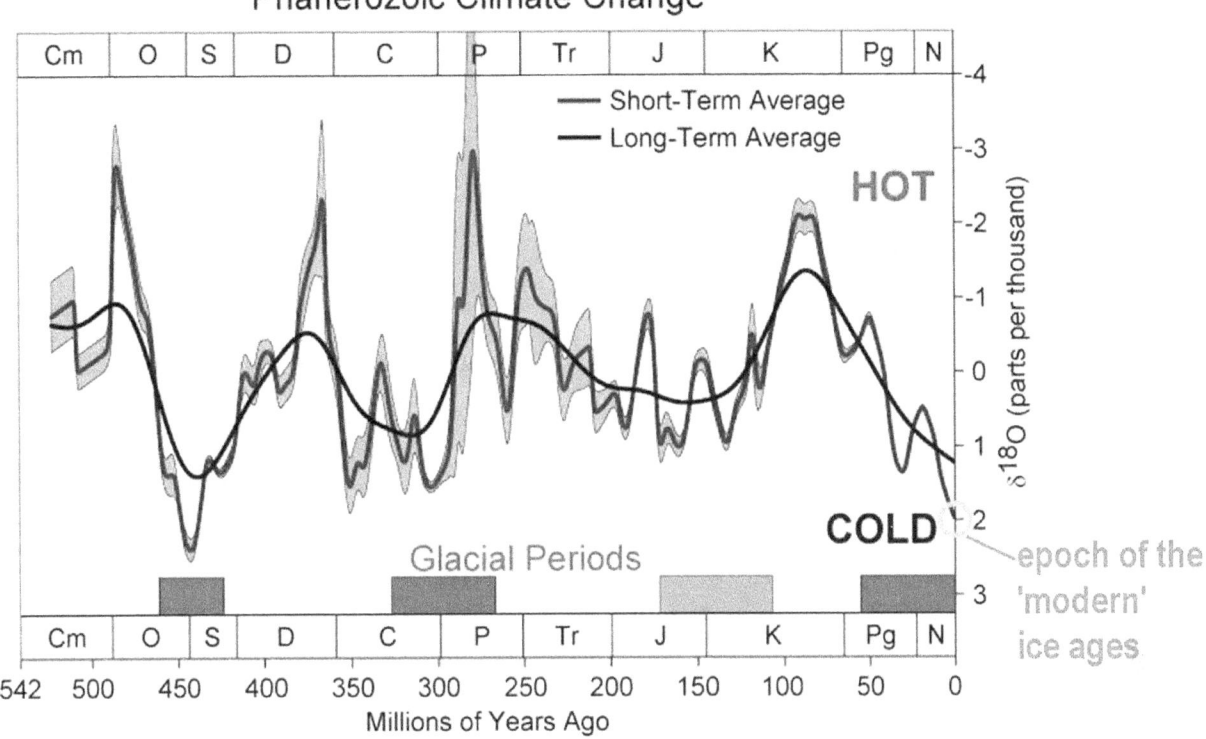

As our Galaxy is presently at its weakest state in 440 million years, our Sun within it operates in a correspondingly weak plasma environment, which is so weak that climate on Earth gets reduced to glaciation conditions under a 70% dimmer and colder Sun for most of its time.

The glaciation conditions get interrupted periodically with warm interglacial periods

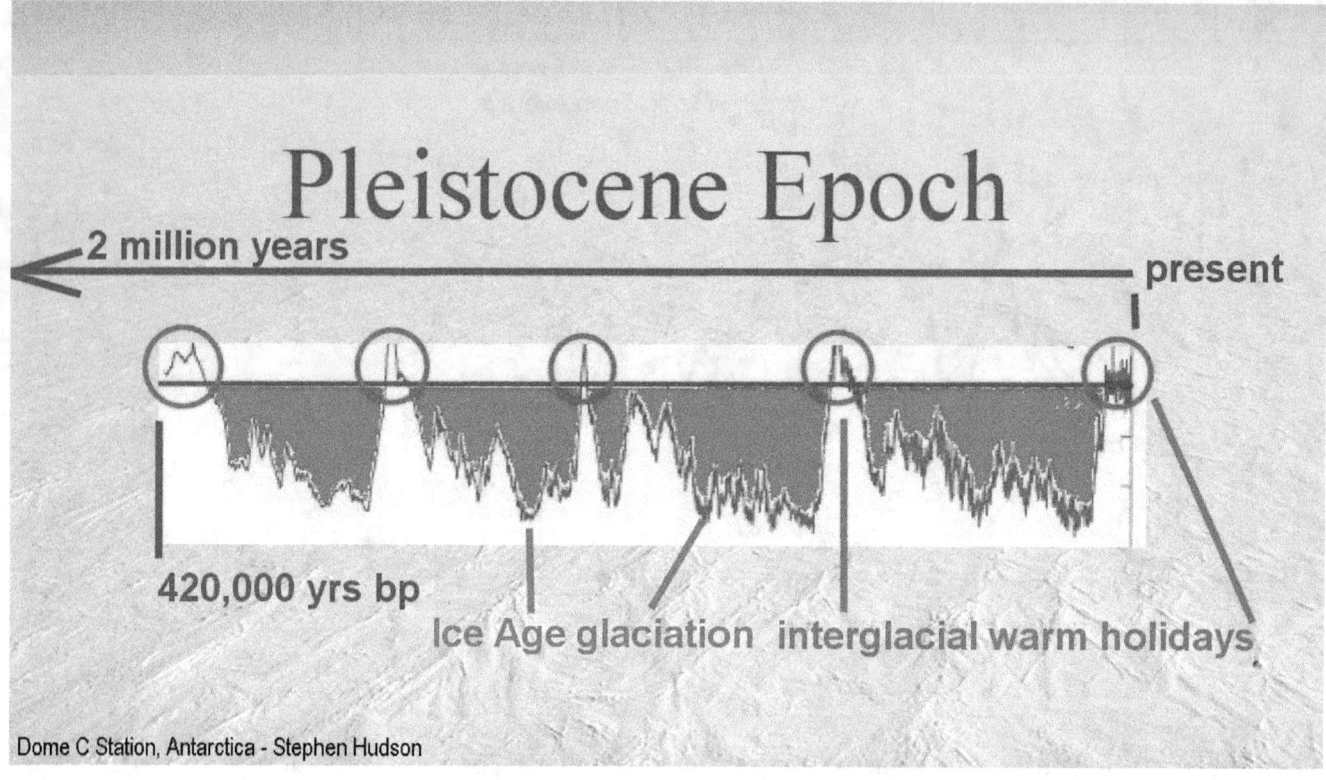

Fortunately for us, the glaciation conditions get interrupted periodically with warm interglacial periods of 12,000 years in duration that occur in roughly 100,000-year intervals. We are presently in a warm period.

The radical difference between glacial conditions and interglacial warm conditions is inherent in the characteristic of the design of the solar system.

The solar system is not as simplistic as I had illustrated earlier

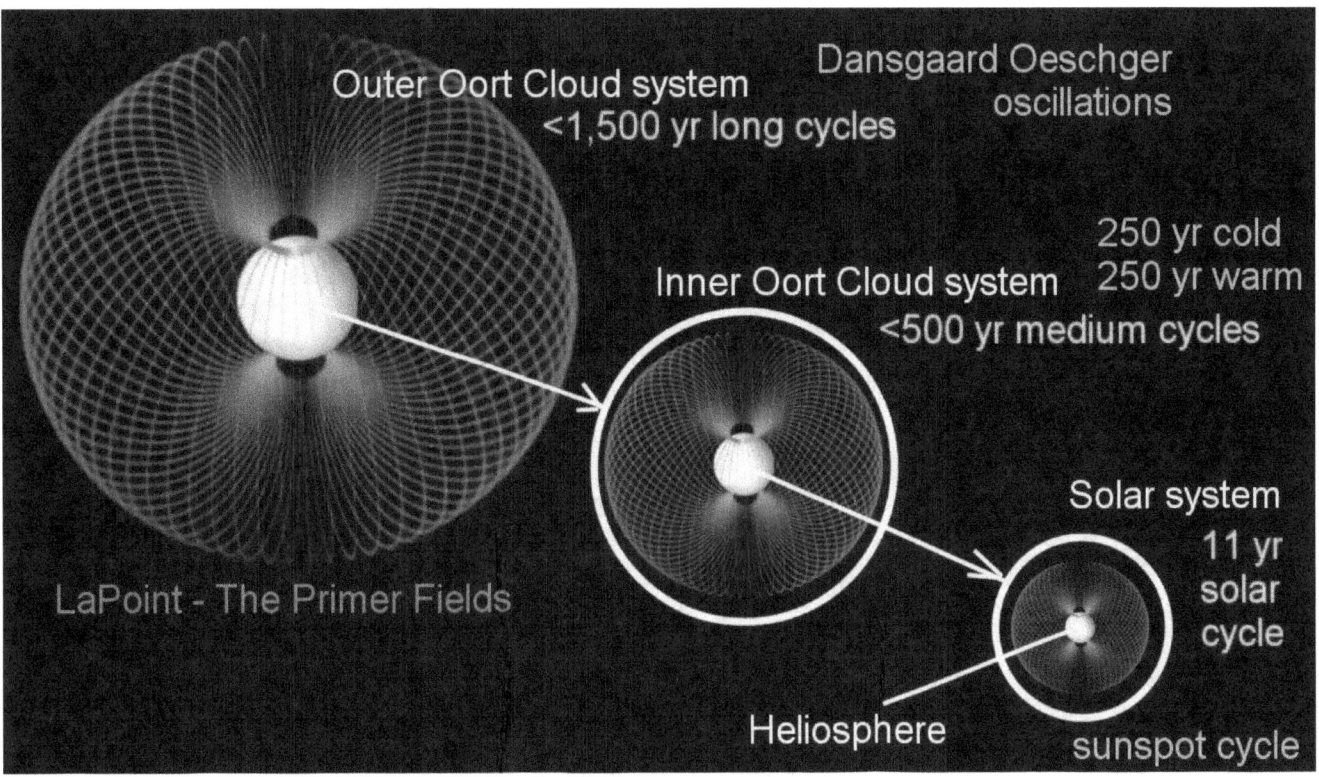

The solar system is not as simplistic as I had illustrated earlier, with a single set of primer fields concentrating the interstellar plasma stream. A number of observed resonance features indicate that our solar system operates as a structure of multiple sets of nested primer fields, with each having its own resonance characteristic.

When the interstellar plasma stream is strong, the flow volume is large enough

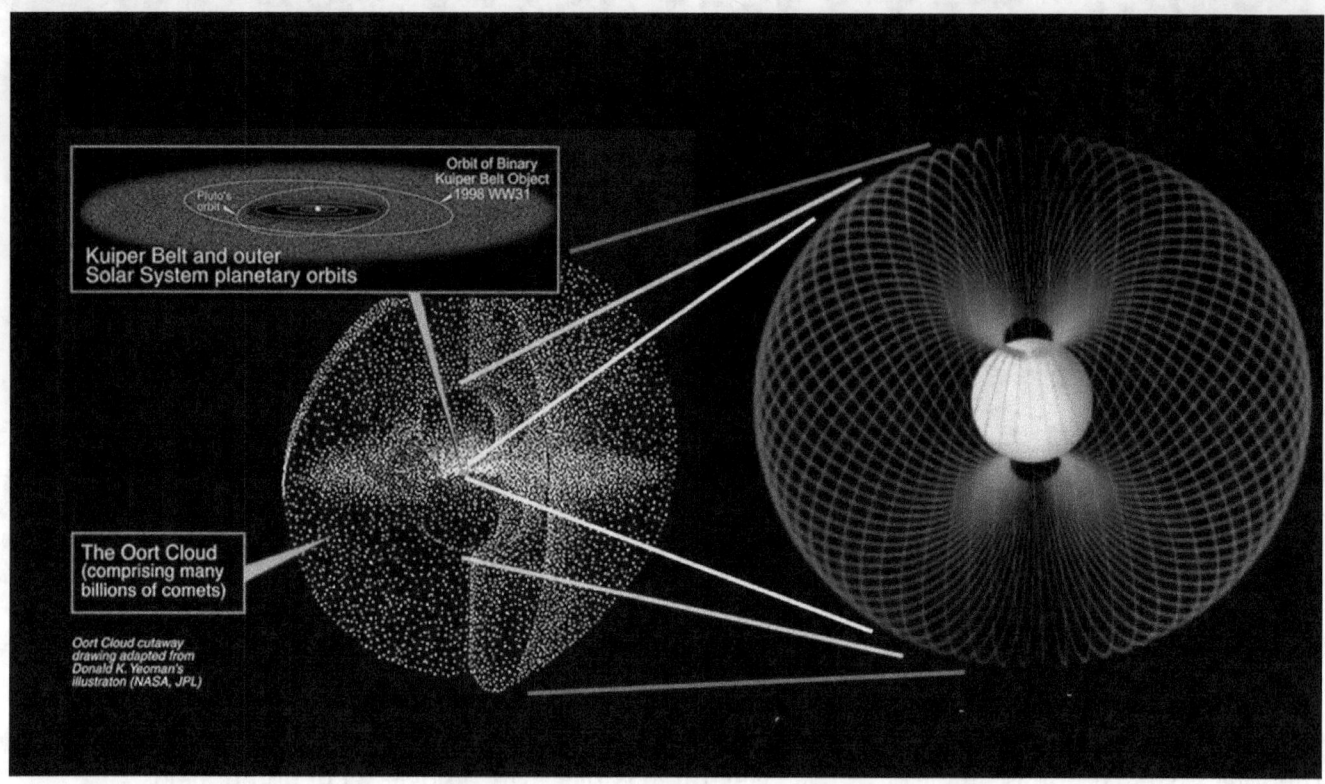

When the interstellar plasma stream is strong, the flow volume is large enough for all nested primer fields to form and operate. In such a case the Sun operates in its high-power mode, whereby interglacial conditions occur on Earth.

Inversely, when the interstellar plasma stream is weak, and the flow volume diminishes below a threshold level for the weakest of the primer fields to form, the plasma density around the Sun becomes radically reduced.

When one set of the primer fields no longer forms

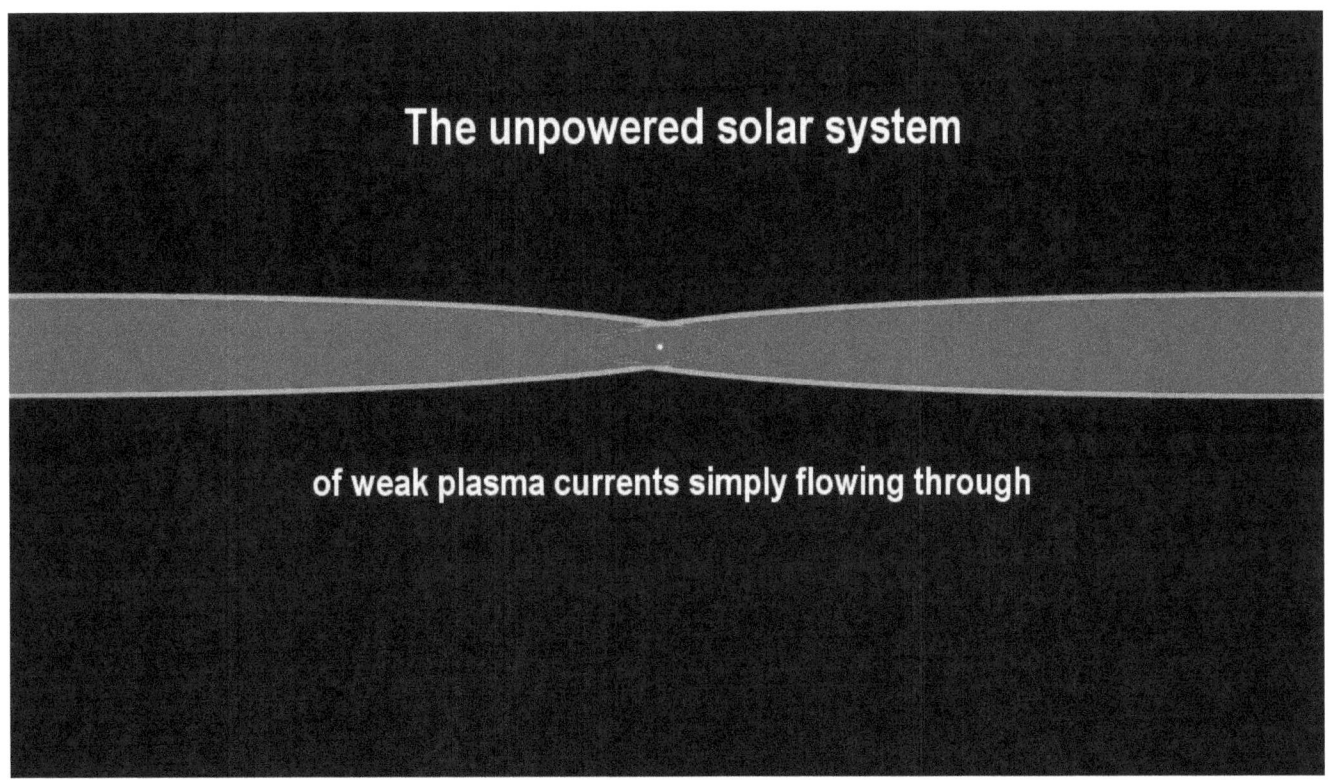

When one set of the primer fields no longer forms, the plasma stream flows around the Sun in a less-focused manner. Under this condition the Sun enters its low-powered state.

At the low-power level the surface temperature of the Sun drops down

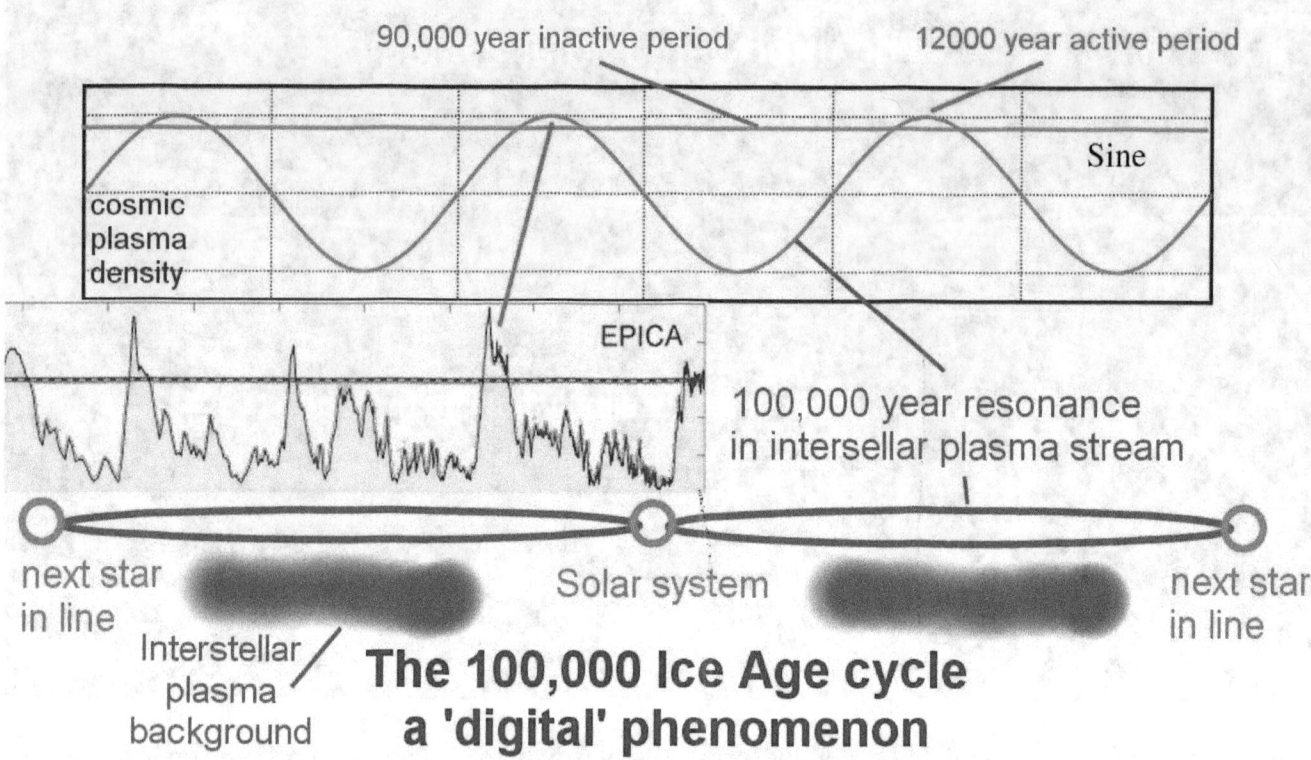

At the low-power level the surface temperature of the Sun drops down from the present 5,800 degrees Kelvin, to possibly 4,000 degrees. At this point the glaciation period begins on Earth. The transition may happen in a very short period of time, possibly in the space of weeks or months. The global temperature will follow the diminishing Sun in rapid succession.

The Ice Age phenomenon for a plasma sun is as simple as that.

For us on Earth, the difference will be felt like a shock

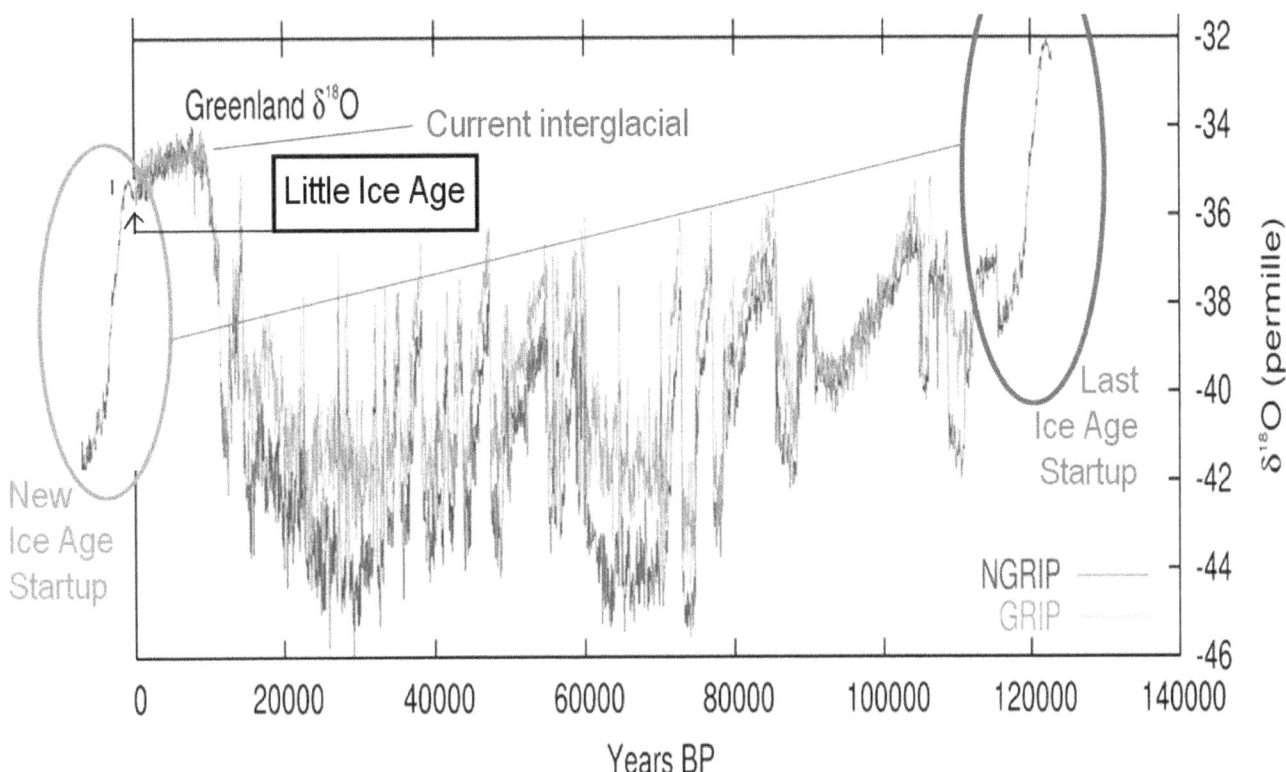

For us on Earth, the difference will be felt like a shock. If one projects forward what ice core records tell us about the previous Ice Age startup, the consequences of this shock will be immense. They promise to be so immense that all areas outside the tropics will become rapidly uninhabitable, and the agricultures become immediately disabled thereby.

➢ **The Sun is in a free fall**

The Sun is in a free fall:

**(no more solar minima)
collapse back into the Ice Age**

The Sun is in a free fall

no more solar minima, but collapse back into the Ice Age.

The timing of the Ice Age start up becomes critical for humanity

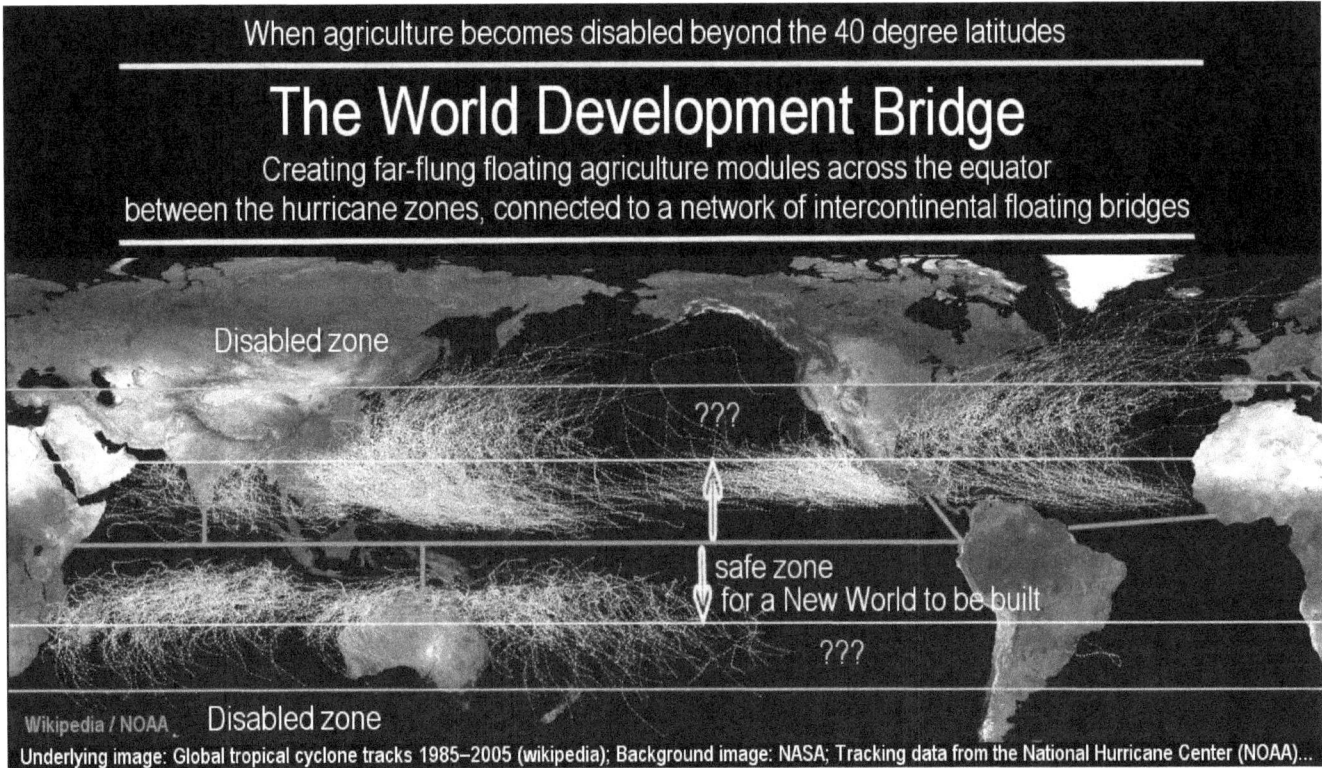

The timing of the Ice Age start up becomes critical for humanity, because the shock can be avoided.

We have the power to prepare ourselves for the Ice Age consequences, before they come upon us, by relocating our living and our agricultures into the tropics. While there is little suitable land available in the tropics, nearly all the new infrastructures and the new cities will need to be located afloat across the equatorial seas, between the hurricane bands. This way we can live securely and richly no matter what the next glaciation period will bring with it. We need to have this ready before the need becomes critical. The exact timing becomes important for this reason. Nor is the timing hard to figure out. The writing is on the wall and is written in huge letters, with the 2050s as the potential timeframe.

The ice core records tell us that our world has been getting progressively colder

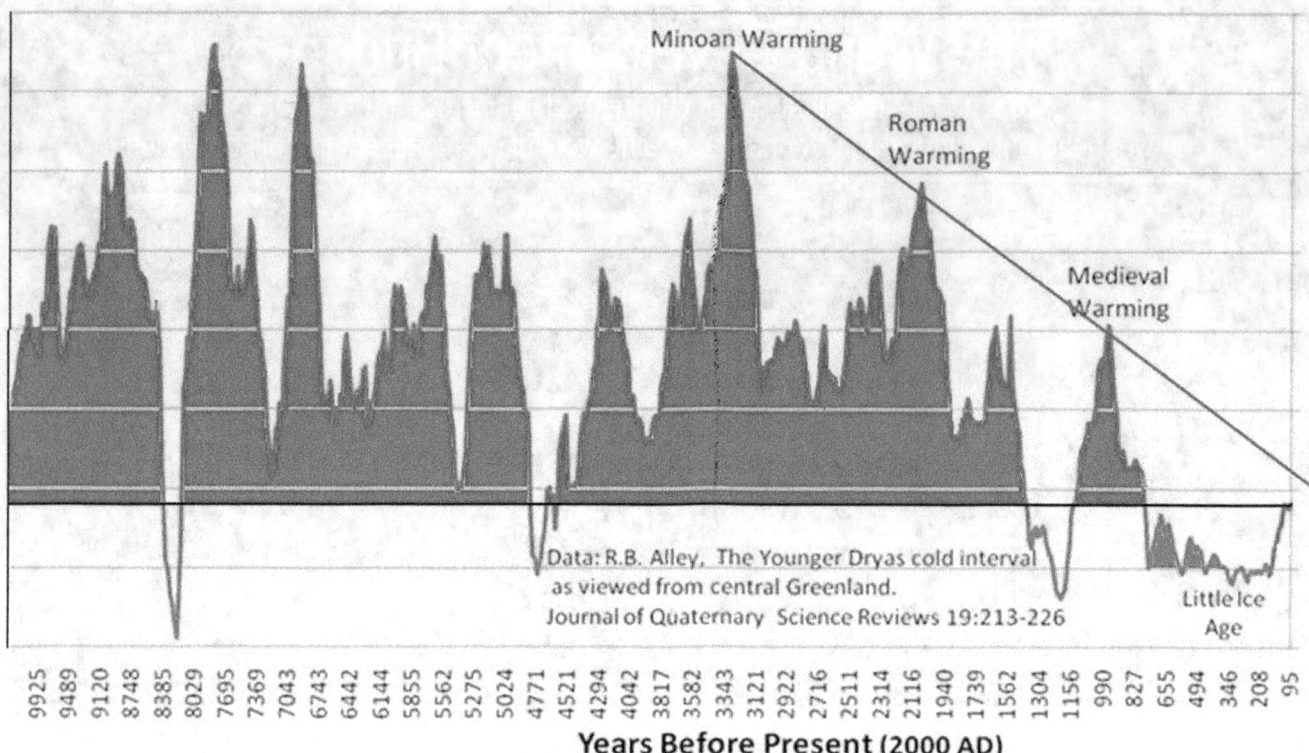

The ice core records tell us that our world has been getting progressively colder for the last 3000 years, with some major warming spikes along the way that are likewise getting smaller. The evidence that pins these fluctuations onto the Sun, has recently been measured in Carbon-14 ratios that stand as a proxy for solar activity.

The plasma fusion reactions on the surface of the Sun

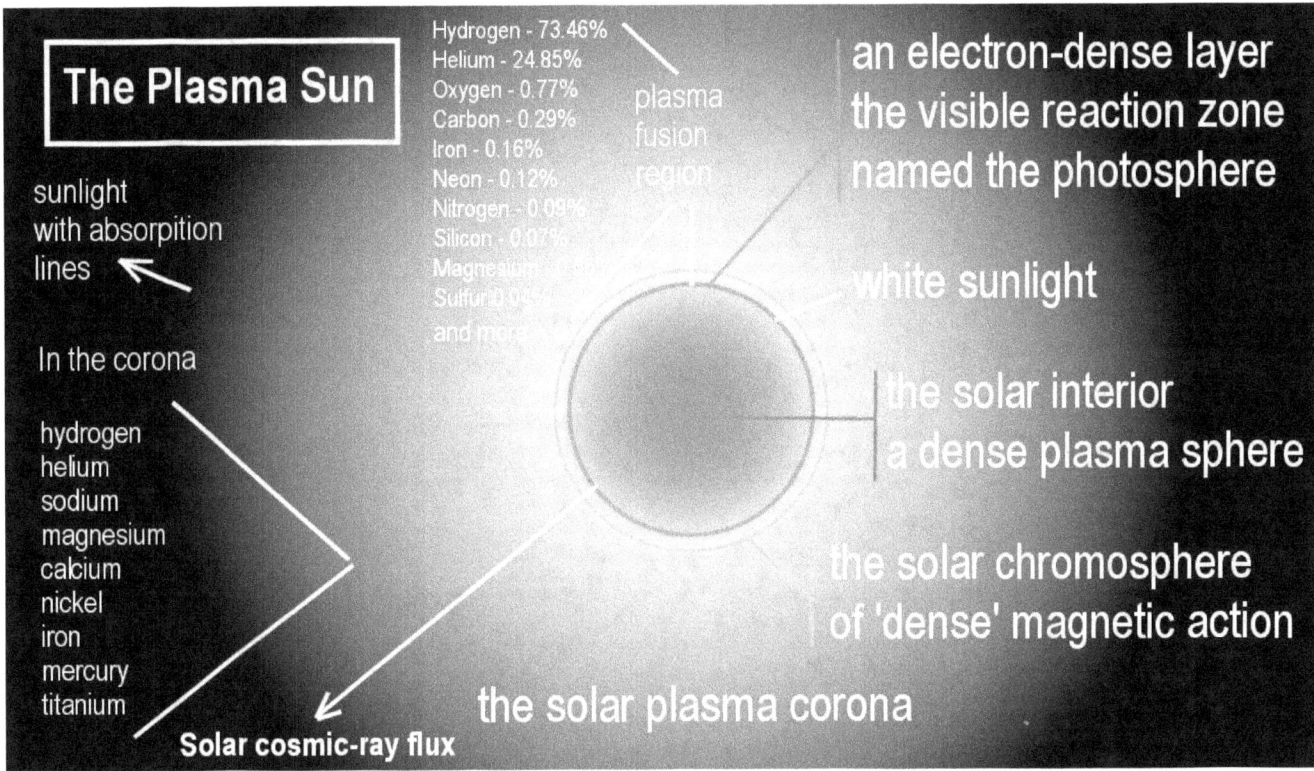

The plasma fusion reactions on the surface of the Sun that synthesize atomic elements, which, energized, give us the sunlight, also produce solar cosmic rays that are single events of plasma particles escaping the reaction chamber in unbound form.

The escaping cosmic rays impact the Earth atmosphere and cause

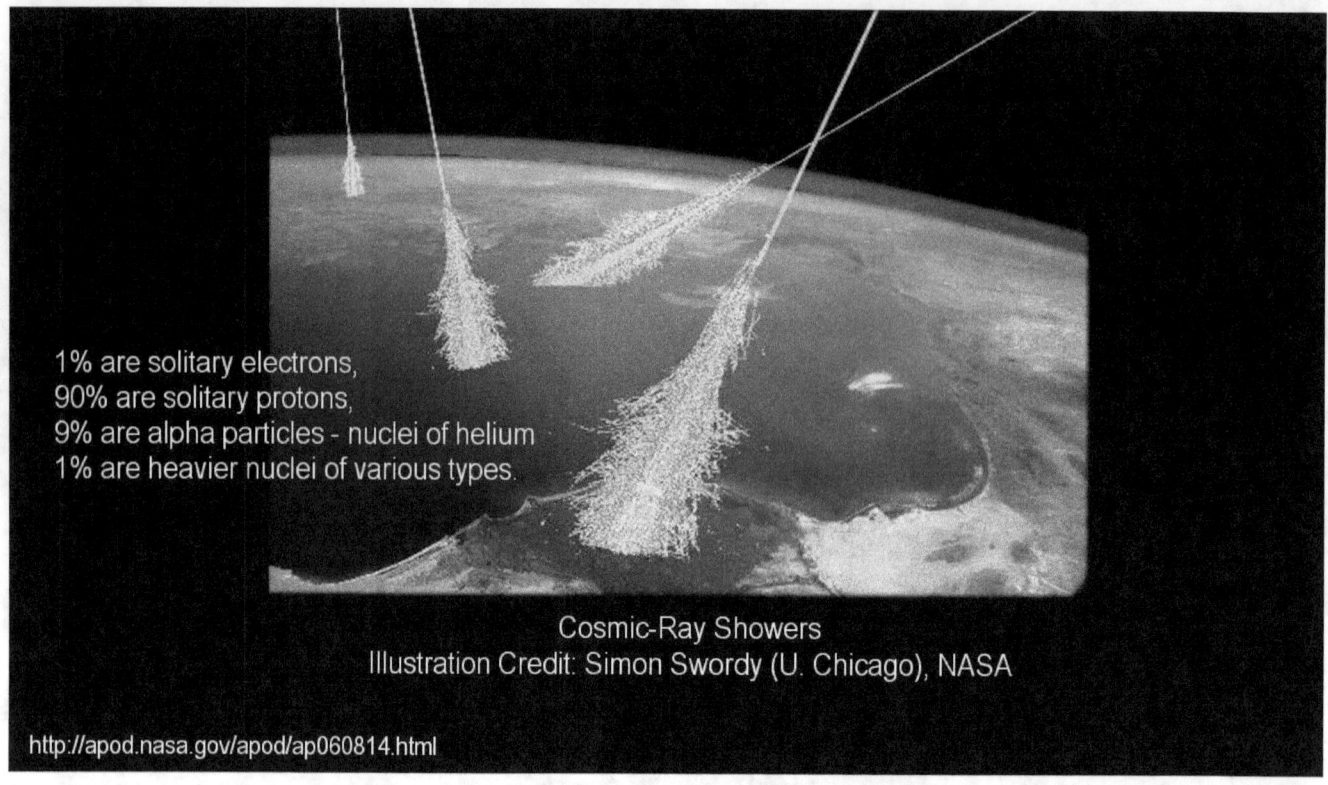

The escaping cosmic rays impact the Earth atmosphere and cause numerous types of reactions. One of these transmutes nitrogen atoms into the radioactive Carbon-14 isotope that can be measured.

The measurements of the ratio in historic samples

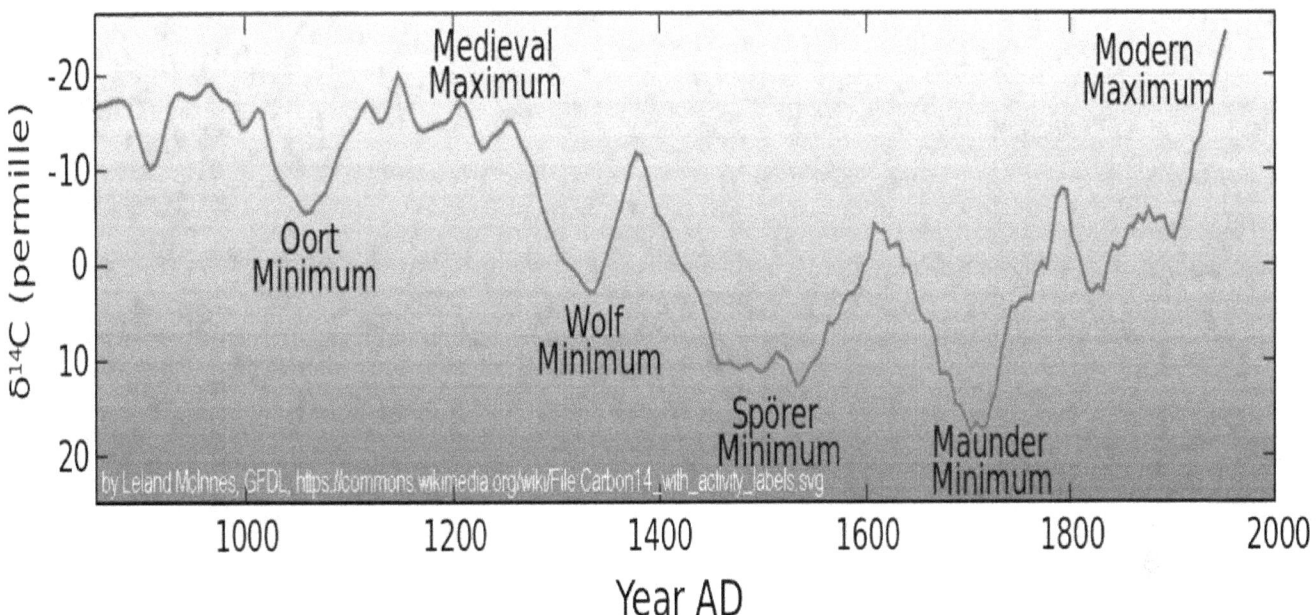

The measurements of the ratio in historic samples becomes a measurable ratio that reflects the intensity of solar activity.

When the interstellar plasma is dense, the plasma sphere around the Sun is dense likewise

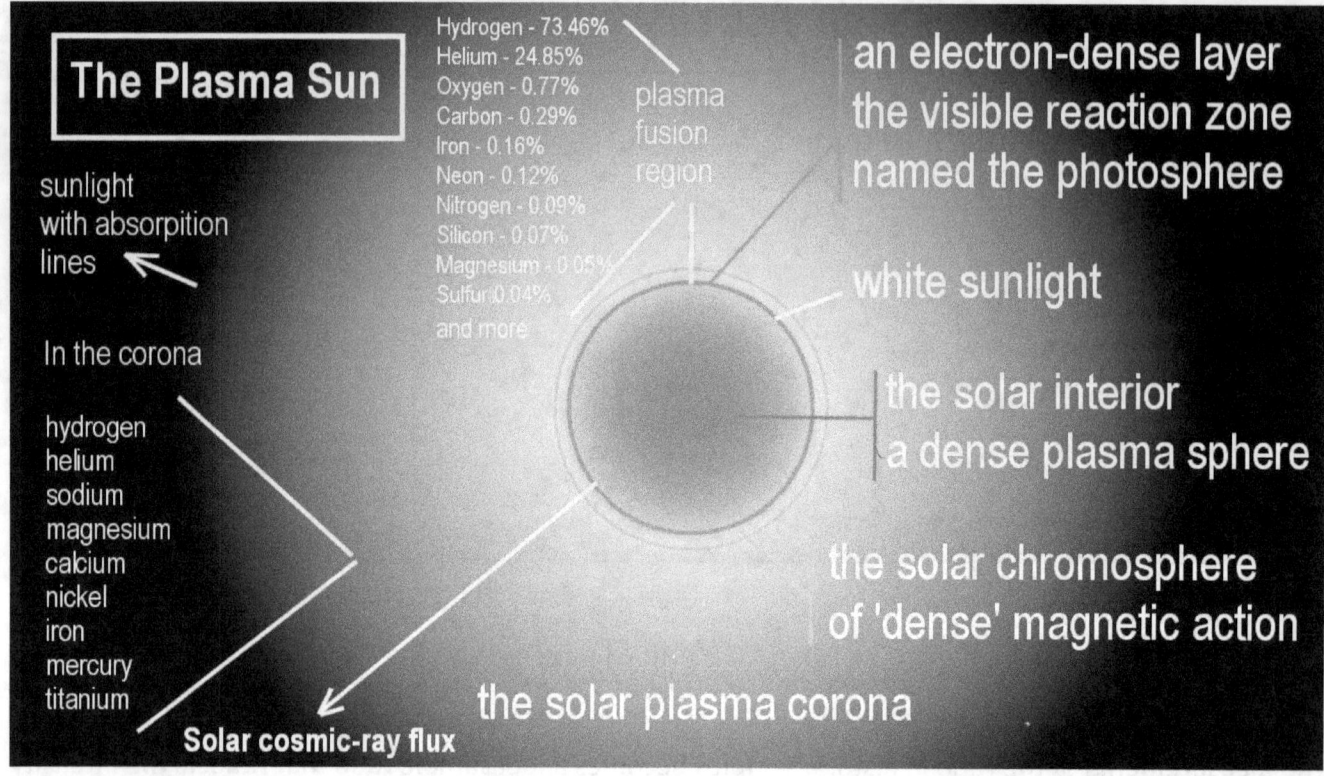

When the interstellar plasma is dense, the plasma sphere around the Sun is dense likewise, which traps most of the solar-cosmic-ray events. This means that fewer cosmic-ray collisions occur in the Earth's atmosphere, which produce lower ratios of Carbon-14.

Inversely, when the interstellar plasma is weak, the solar corona is weak likewise, which enables more cosmic rays to escape, and larger Carbon-14 ratios to be generated.

The changing ratios closely reflect the changing climate on Earth

The resulting trends match the known historic climate trends.

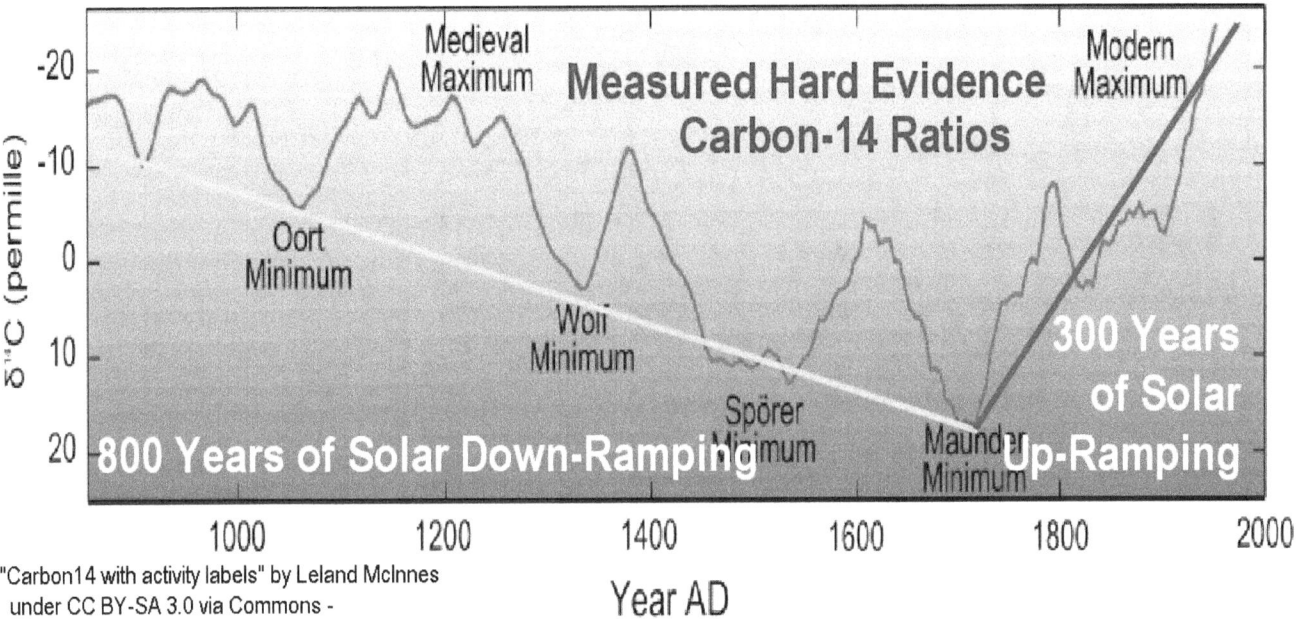

"Carbon14 with activity labels" by Leland McInnes under CC BY-SA 3.0 via Commons -

When one plots these ratios inversely, the resulting trends match the known historic climate trends. The plotted values tell us that the Earth has been in a solar-caused cooling trend for 800 years till the end of the Little Ice Age, after which the Sun recovered sharply till the end of the 1990s.

The interruption of the down-slope in the early 1700s that recovered the Sun and gave us almost 300 years of solar-caused global warming, was evidently caused by one of the long-terms resonance effects that have occurred during the previous glaciation period in intervals of 1470 years, termed the Dangaared Oeschger Oscillations.

The climate spikes of the 3 warming periods that occurred during the last 3,500 years

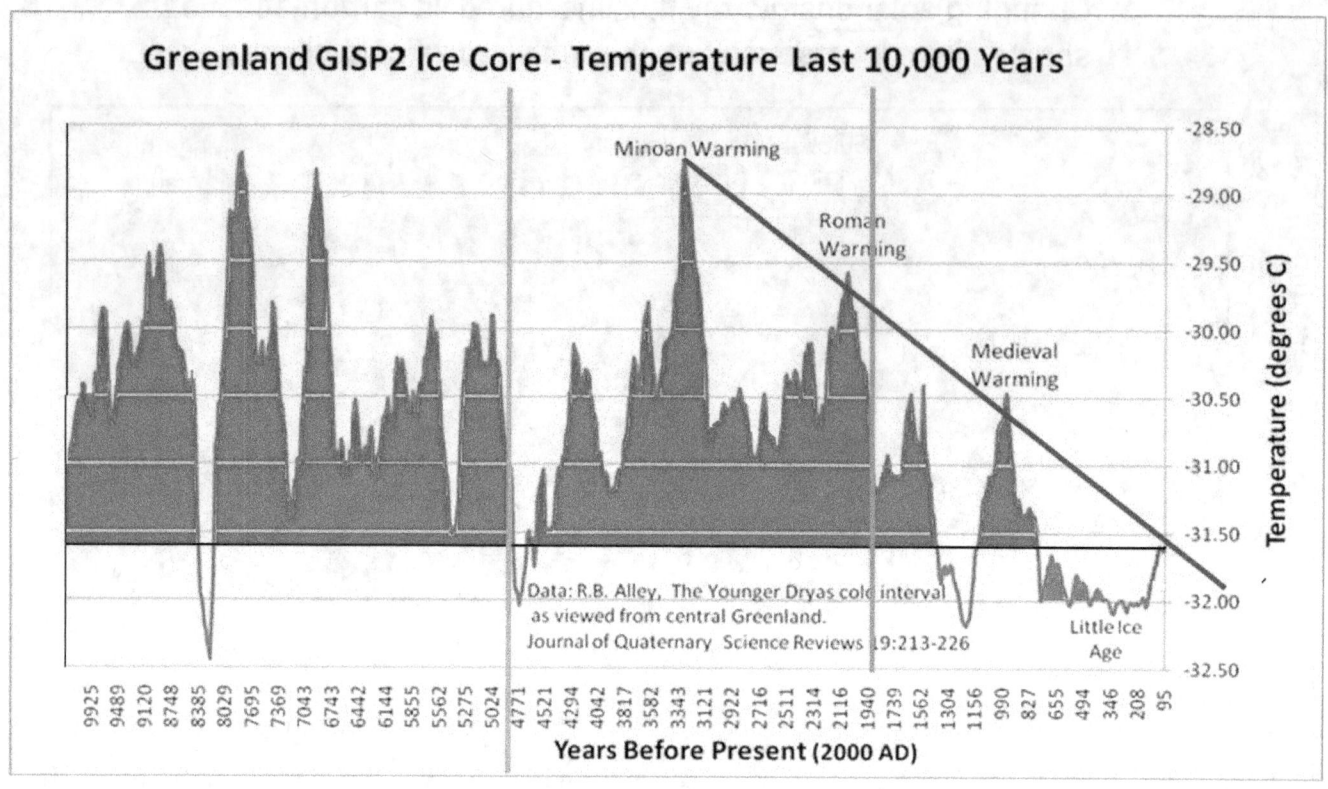

The climate spikes of the 3 warming periods that occurred during the last 3,500 years, may have been related effects.

No matter what had caused the up-ramping of the Sun in the 1700s

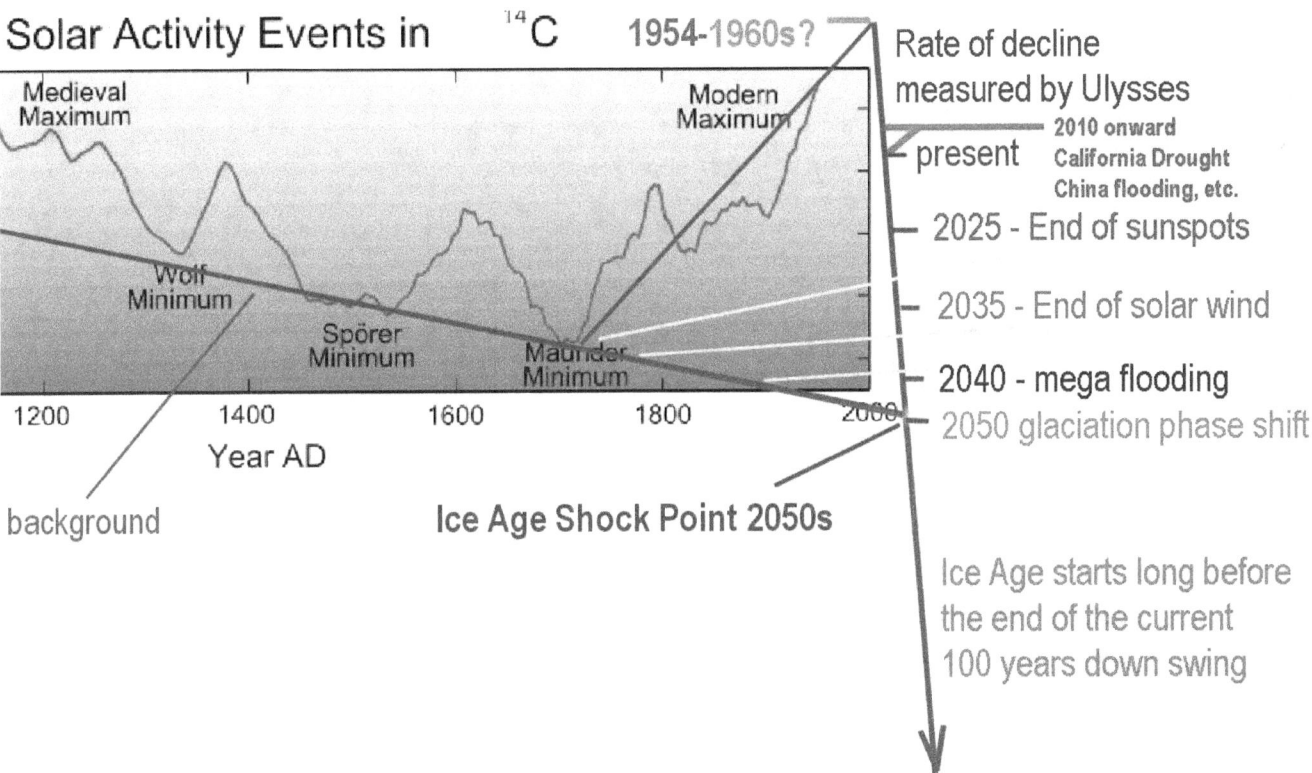

But no matter what had caused the up-ramping of the Sun in the 1700s, the effect of these 1000-year-plus event is over. Its load has been spent. The solar activity is collapsing again.

➢ **Science lets the evidence stand to speak for itself**

Science lets the evidence stand

(to speak for itself)

Science lets the evidence stand to speak for itself.

The rate of the collapse has been measured by the Ulysses spacecraft

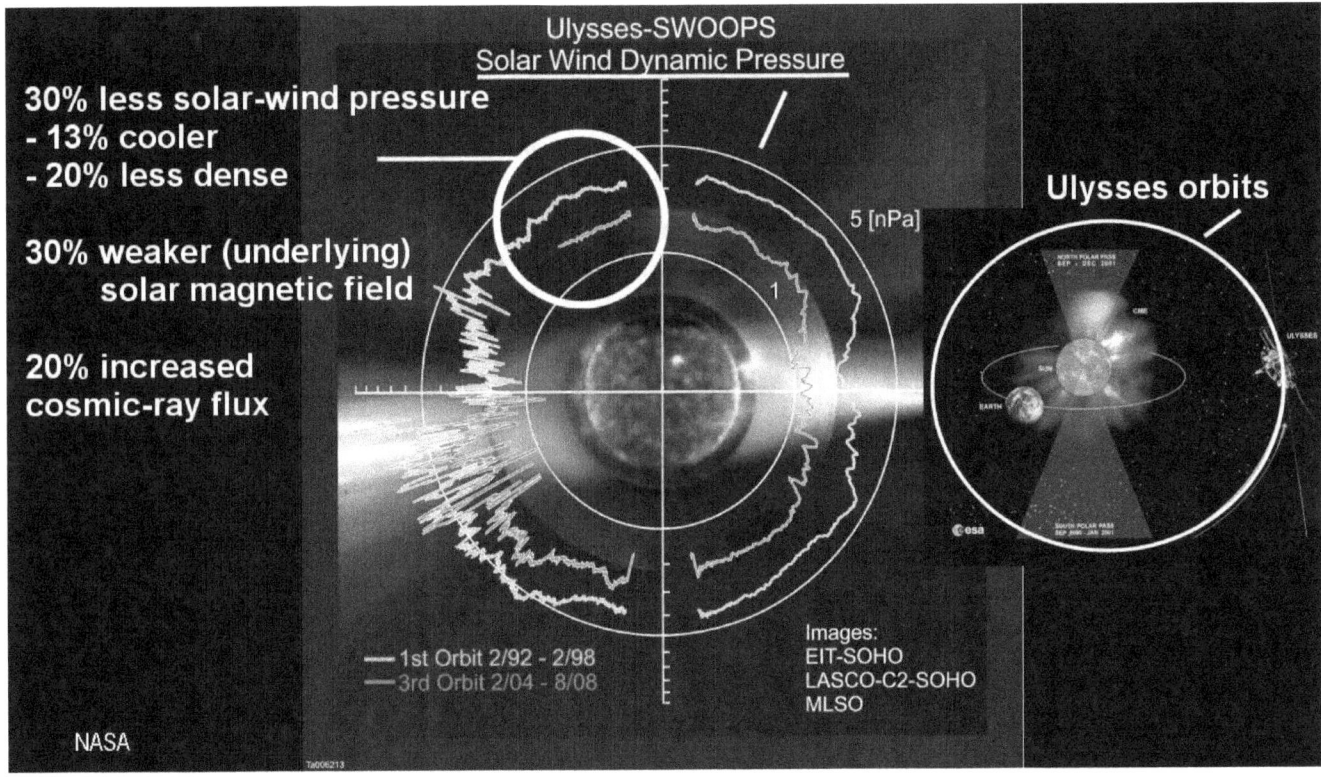

The rate of the collapse has been measured by the Ulysses spacecraft in terms of diminishing solar-wind pressure at a rate of 30% in 10 years. In cosmic terms, this steep rate of decline is akin to a free-fall collapse. The effects of this steep collapse are visible everywhere.

With the solar cosmic-ray flux sharply increasing, under the weakening Sun

With the solar cosmic-ray flux sharply increasing, under the weakening Sun, we see cloudiness increasing on Earth at an ever-faster rate. Cosmic-ray flux has an ionizing effect in the atmosphere which enhances cloud nucleation up to 100-fold. The resulting increased cloudiness reflects more of the incoming sunlight and solar energy back into space, which becomes lost to us and makes the Earth colder. Increased cloud nucleation, and its faster rain-out, causes increased flooding and also increased droughts by reducing the water transport distance in the clouds. We see all of these effects now occurring evermore.

The rate of solar collapse that the Ulysses spacecraft saw still continues

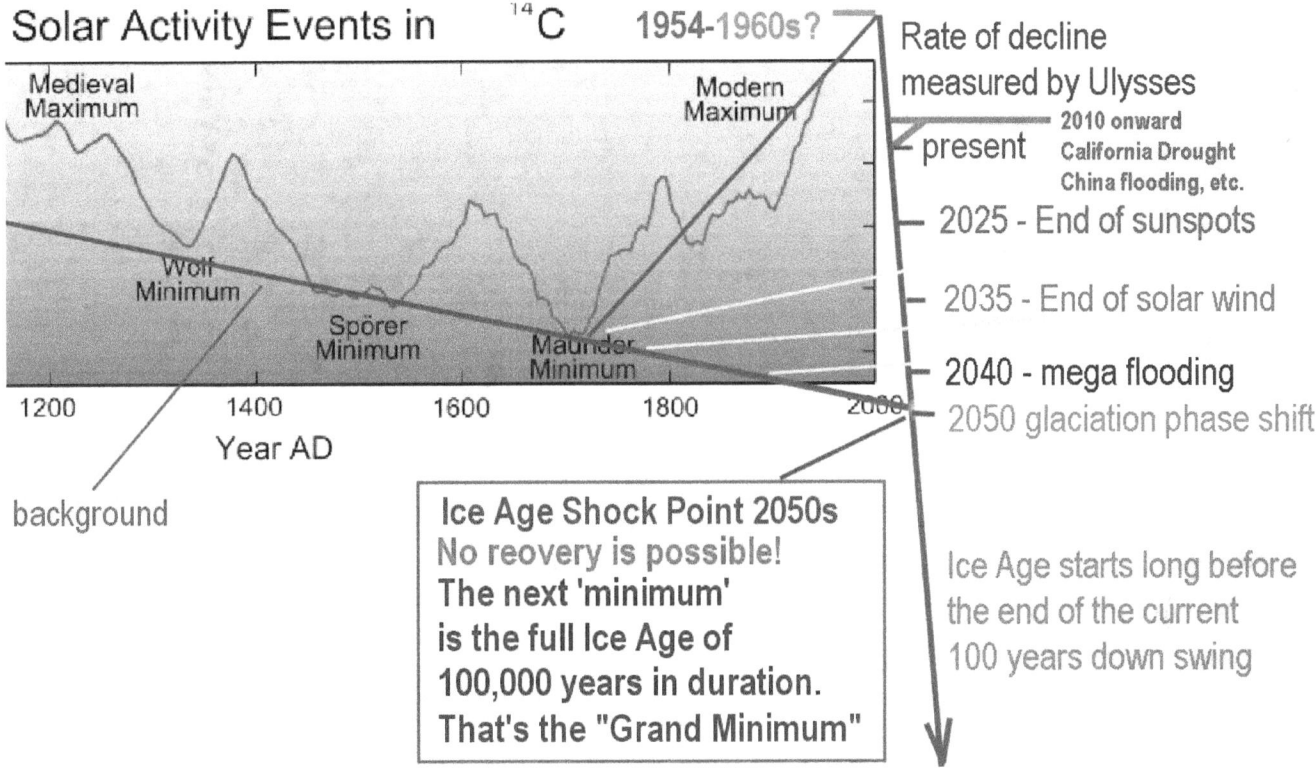

The rate of solar collapse that the Ulysses spacecraft saw only the beginning of, still continues.

The measured rate of collapse promises the end of the sunspot cycles in the 2020s and the end of the solar wind in the 2030s, after which the solar surface temperature will begin to diminish.

Along this path the threshold becomes crossed when the innermost primer fields can no longer be maintained and the full Ice Age begins.

The event could happen in the 2040s and onward into the 2050s. However, we should also be extremely worried from the 2030s on, because when the solar wind diminishes to near zero, the Sun looses one of its mechanisms for purging the fusion cells of their synthesized products. Without the solar wind the fusion cells tend to clog up, which diminishes the fusion process further. The clogging up may radically accelerate the collapse process towards the Ice Age start-up phase shift.

The uncertainty in timing makes the rapid building of the necessary infrastructures in preparation for the coming Ice Age phase shift, evermore urgent.

We have laboratory-developed proof that when a fusion cell is clogged up

We have laboratory-developed proof that when a fusion cell is clogged up with its own fusion product, the fusion reaction cannot be long maintained. The longest duration nuclear fusion reaction lasted one second, a kind of world record. This has been achieved with this largest European experimental reactor. It clogged itself up after one second, by which the reaction stops.

Our Sun may get into this type of critical situation in the 2030s when the solar wind dies down to zero. The phase-shift to the Sun's lower-energy state, with which the Ice Age begins anew, might be triggered by the same type of process that blows out the fusion reaction in the laboratory. The clogging up is one of the stoppers that makes the current concept of nuclear-fusion power production an unrealizable dream.

The laboratory experiment also proves that the Empire Sun is an impossible dream

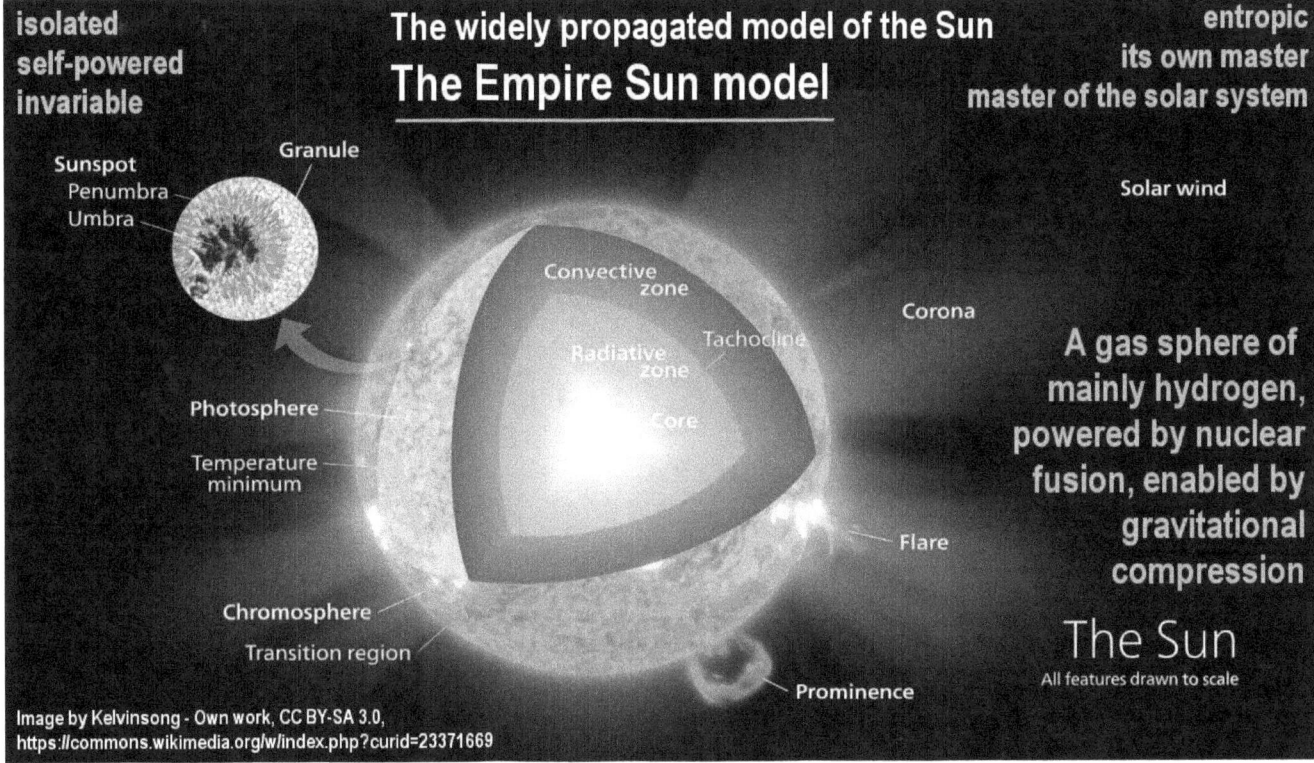

The laboratory experiment also proves that the Empire Sun as a gas sphere, that's supposedly powered by hydrogen being fused into helium at its core, is inherently an impossible dream, as the heavy fusion product, which by its weight would remain in the core, would clog up the hydrogen fuel and extinguish the fusion reaction. The danger that the Empire Sun imposes on humanity, as an impossible dream, is that the dreaming clogs up the landscape of science and thereby prevents the recognition of the real nature of the Sun and its variable quality by its dependence on external energy supplies that are presently diminishing towards such weak conditions that the Ice Age phase shift looms in the near future.

If this clog in scientific recognition cannot be cleared, NO preparation for the coming Ice Age will be made, as is presently the case, and doom awaits humanity.

We may see the building of the floating infrastructures to begin soon

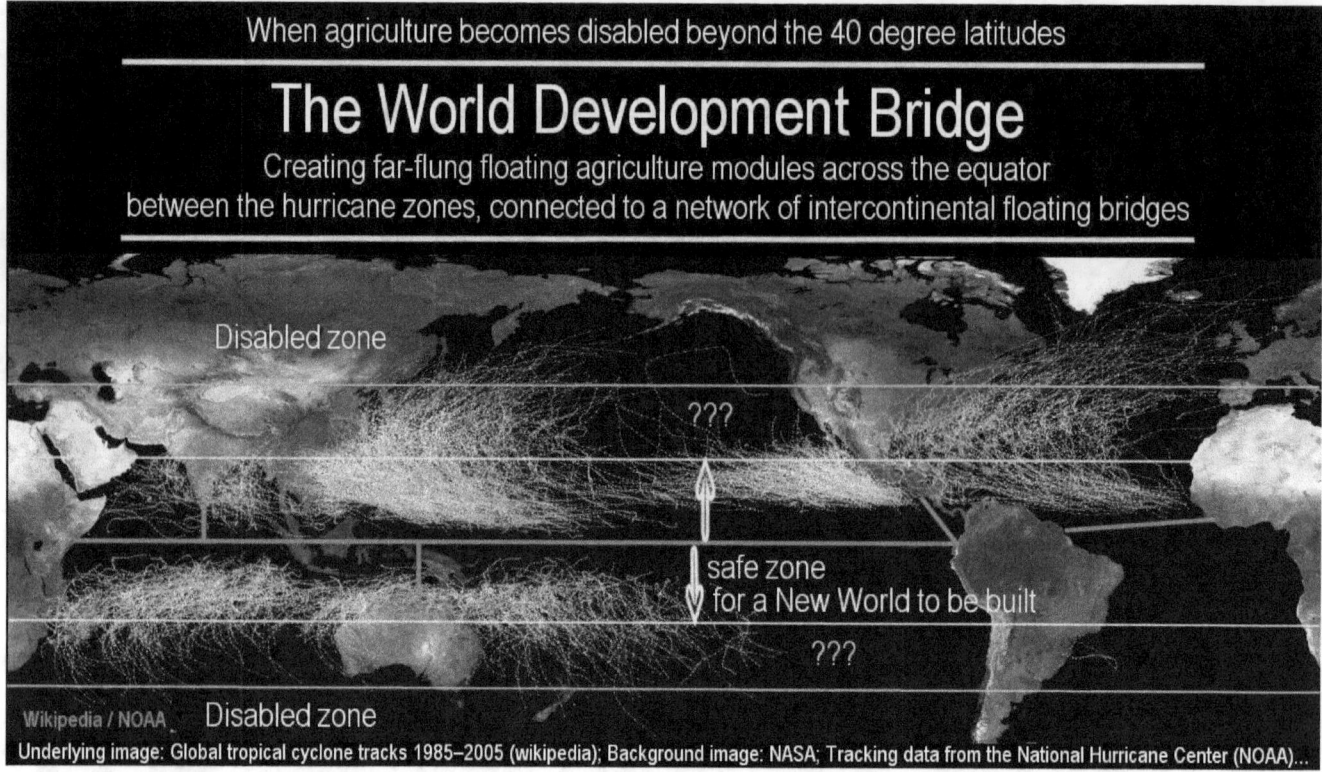

Fortunately we may see the building of the floating infrastructures to begin soon, in spite of it all, because as the Sun weakens towards the big Ice Age phase shift, the solar cosmic-ray flux emitted by the Sun will so dramatically increase, that the resulting increased flooding and drought conditions, and possibly also earthquakes, will cause a movement to happen onto the sea in the form of floating infrastructures being built. It is inherently more efficient to escape epic-style flooding and droughts, than it would be to prevent them. Infrastructures that are located afloat on the seas are immune from flooding. Flooding is not possible there. Earthquakes have no effect there. Droughts cannot happen there either, where freshwater supplies can be drawn from the oceans by deep ocean reverse osmosis desalination.

In order to make the full relocation of humanity into the tropics possible

And in order to make the full relocation of humanity into the tropics possible, that the now changing climate will soon inspire, the 6,000 new cities that are required for this will be built. They will be built with automated industrial processes so that they can be provided to one-another for free as a critical investment by society into itself.

All of this is possible. The materials exist. The technologies exist. The energy resources exist. The only element that does not exist at the present time is the willingness in society to get the process started that enables its future existence in the changing world before the onrushing Ice Age will actually be recognized.

One of the chief blocking factors that prevents the Ice Age recognition

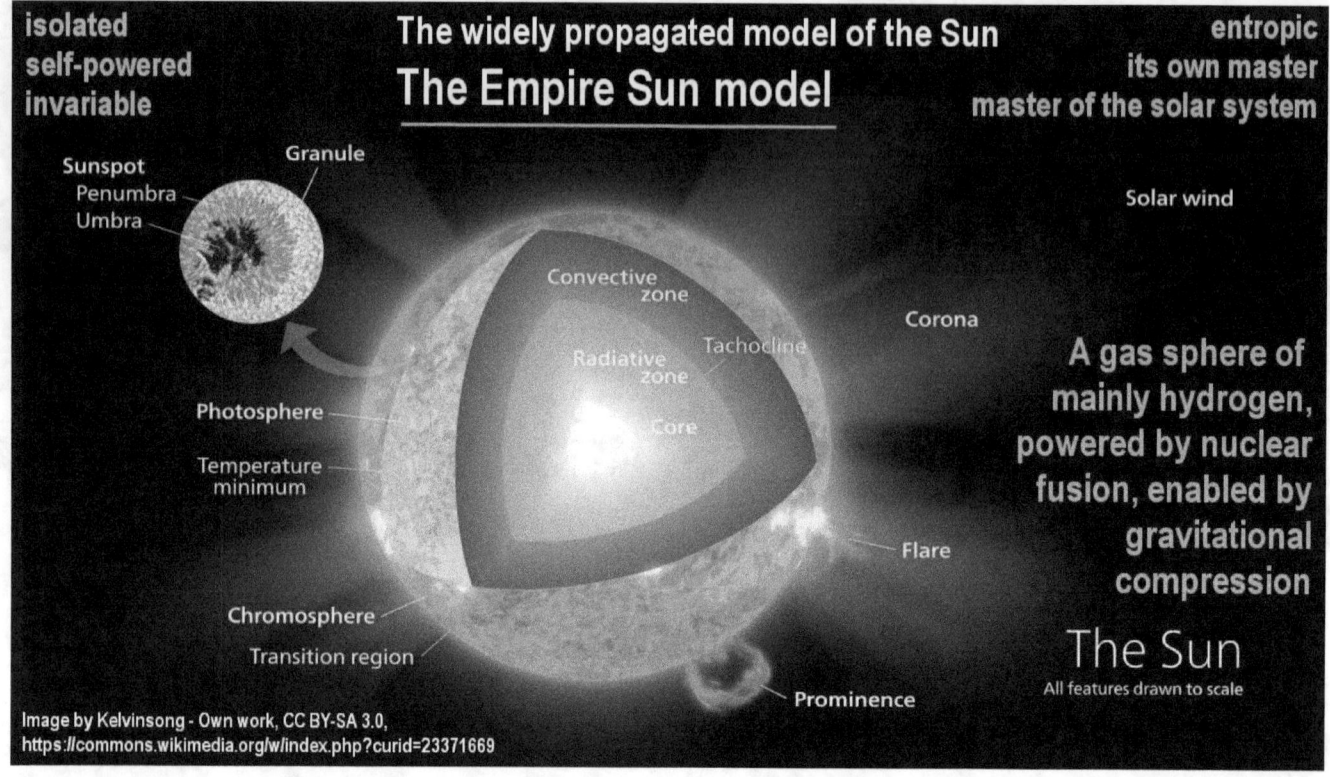

One of the chief blocking factors that prevents the Ice Age recognition at the present time, that blocks the fully focused preparations for it, is society's tenacious clinging to the Empire Sun, the hydrogen-sun theory, of which nothing is real - which mirrors the nature of empire itself where nothing is real either.

Obviously the breakout to sanity needs to begin soon, with a step away from the doctrinal dictates that presently smother the scientific recognition of humanity and assures society's doom.

The poster of the Empire Sun is that of an invariable Sun, which, by long schooling is reflected in people's clinging to the school-taught doctrine. Just as the victor in war writes the history books, so, what is science and truth, is written in the same manner.

Under the invariable Sun all climate changes are said to be necessarily man-made

Under the paraded poster of the invariable Sun all climate changes are said to be necessarily man-made, and ice ages are said to be not possible except by long-term orbital variations over the span of thousands of years that are still far, far distant in the future, too distant for anyone's concern.

The orbital cycles theory, termed the Milankovitch Cycles theory

The orbital cycles theory, termed the Milankovitch Cycles theory, ironically is not physically a possible cause for ice ages, for the simple fact that the solar energy received from the Sun remains the same regardless of cyclical variations of the tilt of the spin axis of the Earth and the eccentricity of its orbit around the Sun.

The Milankovitch Cycles theory has been attached to the Empire Sun doctrine

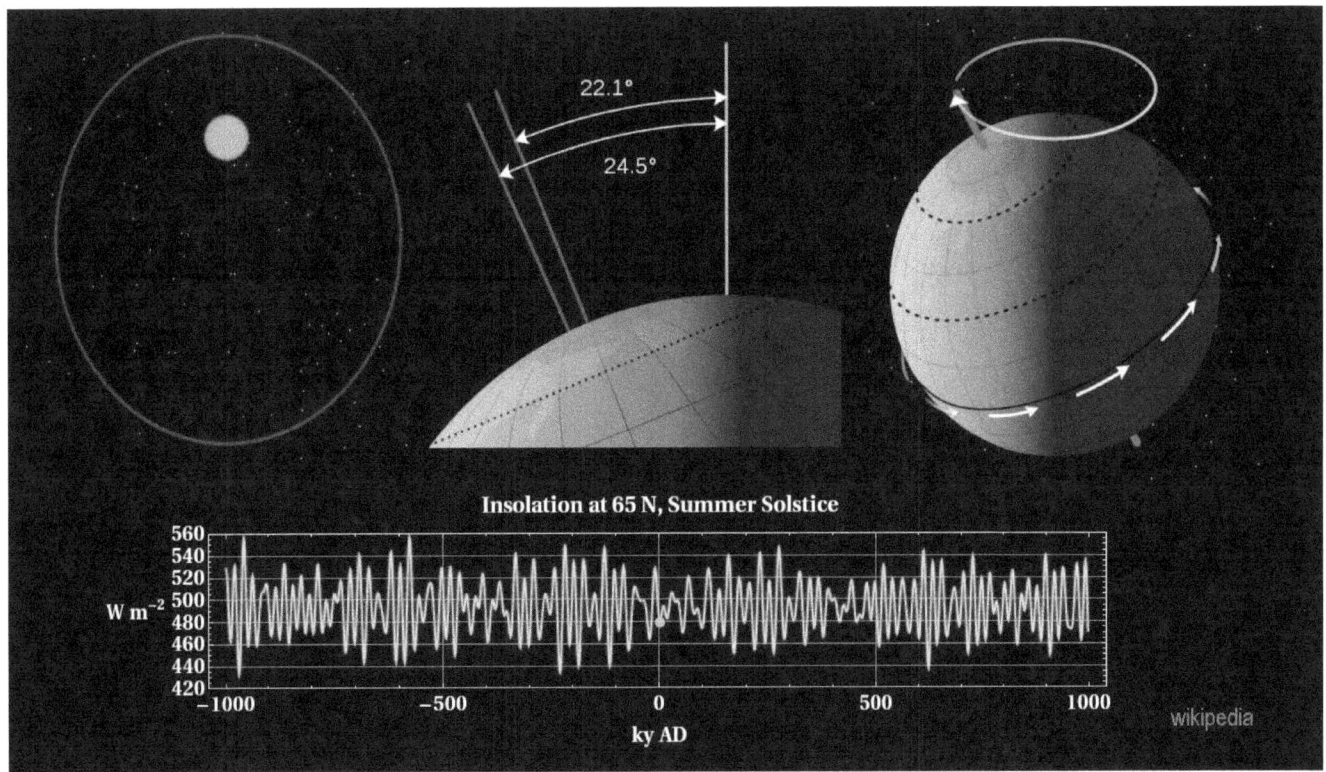

The Milankovitch Cycles theory has been attached to the Empire Sun doctrine, as a doctrine for the ice ages. It is so far off the track from discovered reality that Johannes Kepler had already disproved the theory in principle back in the 1600s, long before the empire-related theory was invented. The theory that has become a doctrine, like the man-made climate change doctrine, has no foundation to stand on. Both are evidently both deployed for their obscuring characteristic that prevents the recognition of the Ice Age dynamics, and thereby prevents humanity from preparing its world for the near Ice Age transition in the 2050s.

If the Empire Sun is retained

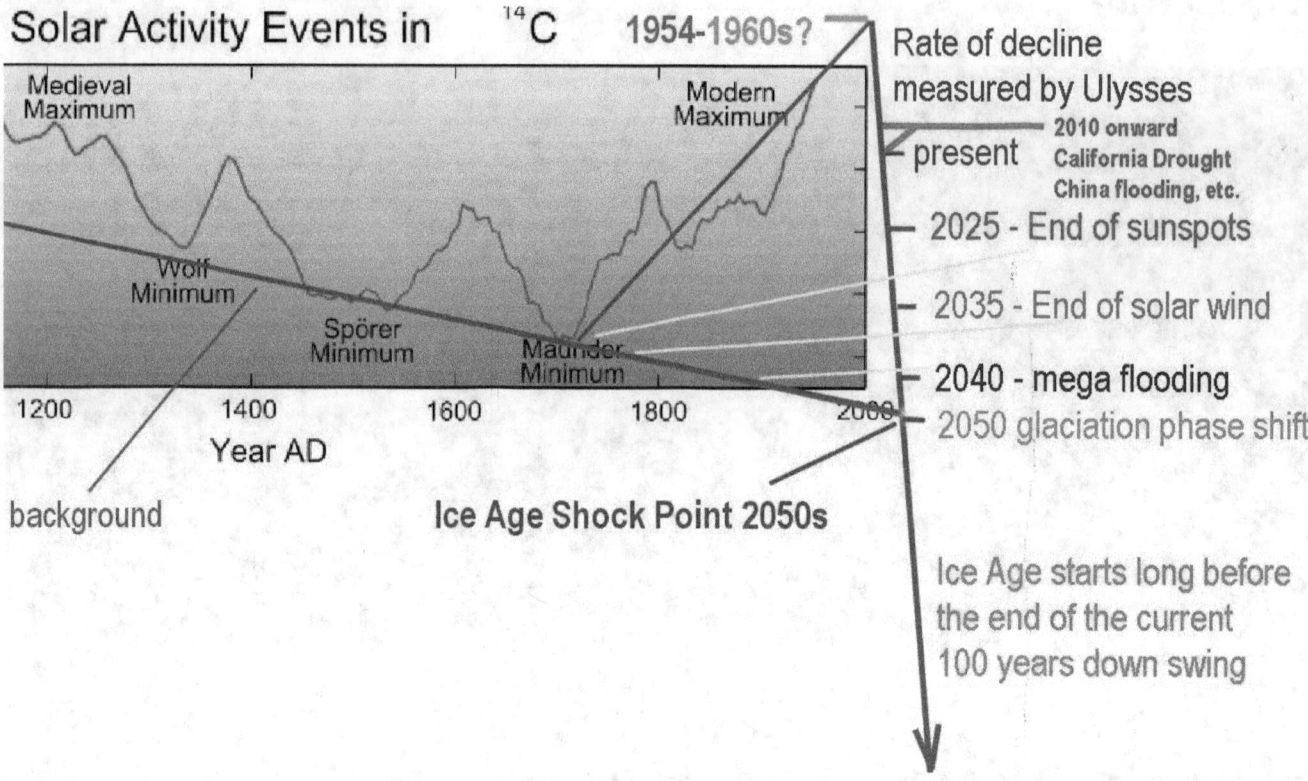

If this imposed obscurity by empire is not lifted, and the Empire Sun is retained, which stands at the center of all that is false in climate science, the inevitable consequence will be the most radical depopulation of the planet that has ever occurred in the entire history of civilization. the astrophysical collapse is in progress. If the food-production infrastructures that qualify for Ice Age conditions are not created before the Ice Age starts, all but a few of humanity will starve to death.

That's the significance of the Empire Sun poster

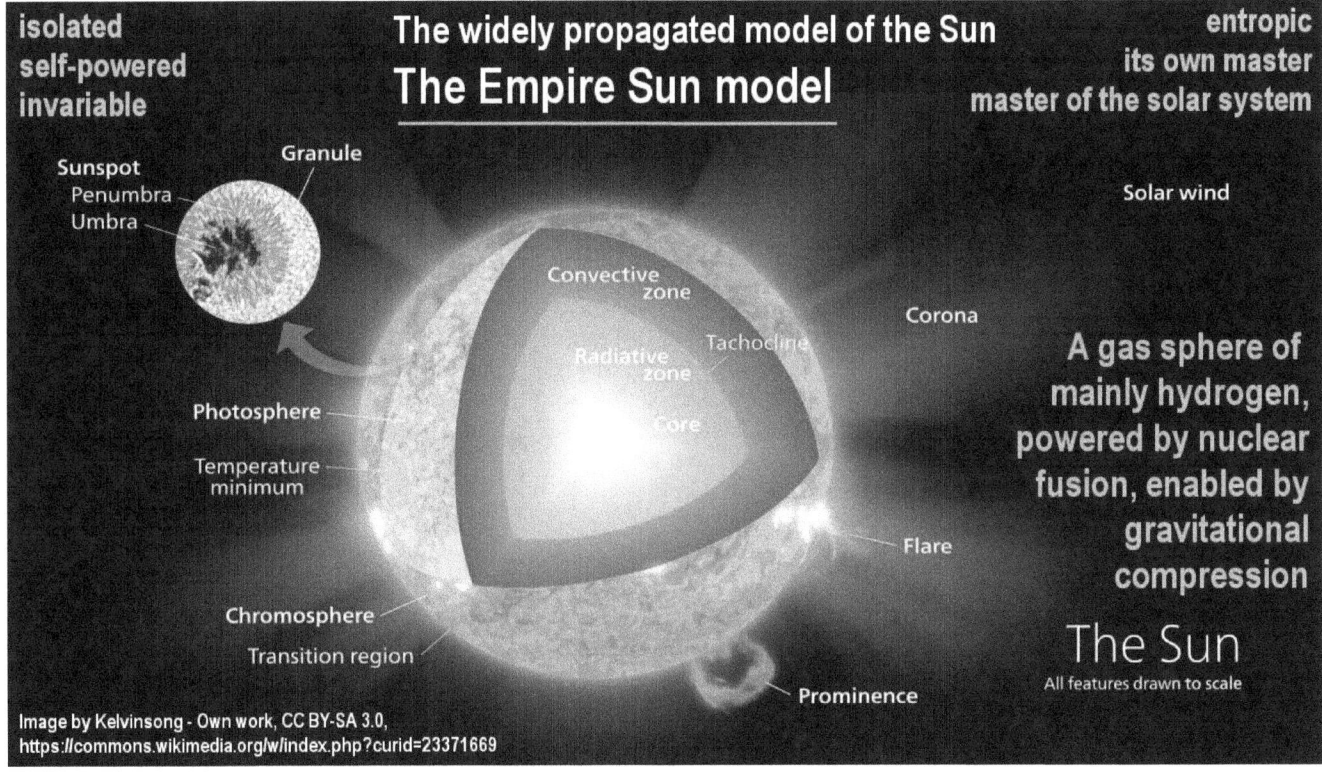

That's the significance of the Empire Sun poster.

The resulting radical depopulation that the false poster sets the stage for

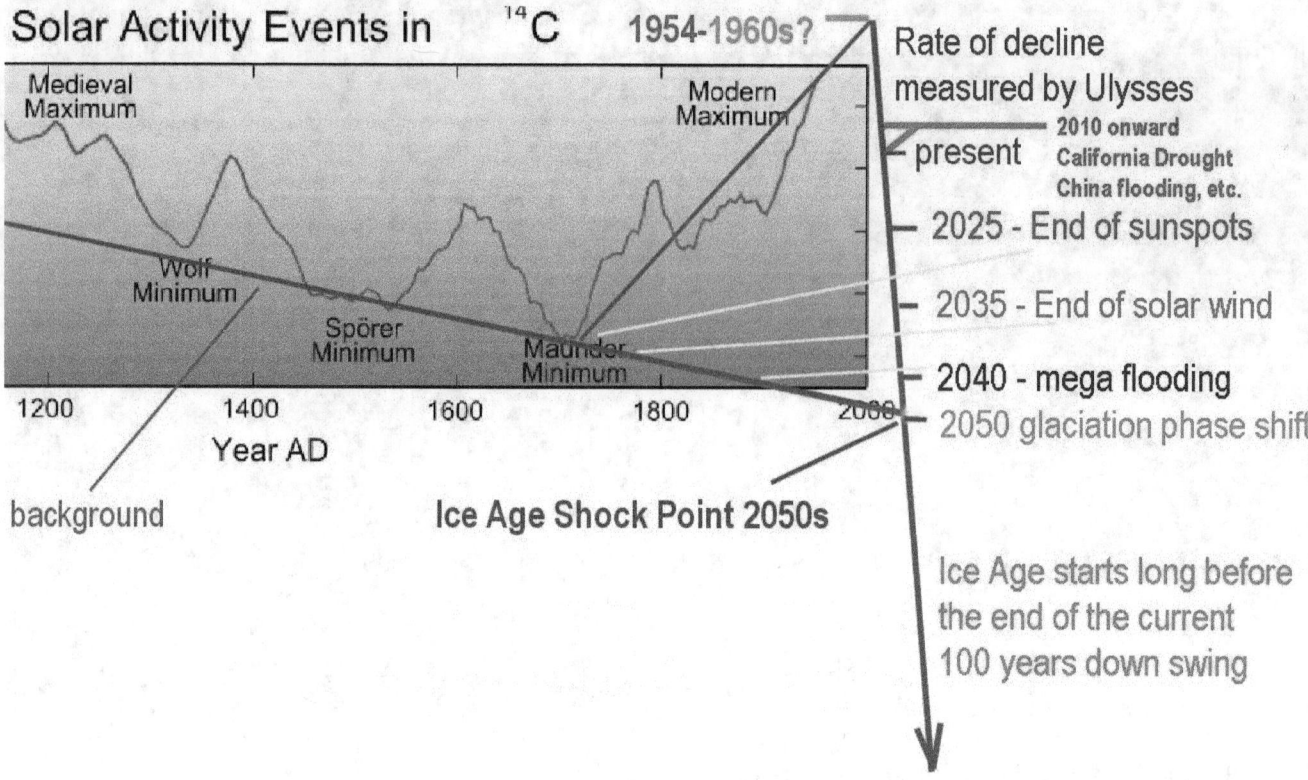

The resulting radical depopulation that the false poster sets the stage for, would be the very effect that the masters of empire have long desired. According to their own saying, the current policy for human depopulation is to eliminate six billion people from the face of the planet as fast as possible. This has been announced many times. The Manmade Climate Change doctrine is a part of the depopulation package, a package of mass genocide. In principle, mass genocide has been an inherent part of the policy of the system of empire for as long as empires of numerous types and names have disgraced the face of humanity from ancient times onward.

Part 3: Dawn of freedom, from empire

- **Love versus Empire - the beginning of the end of Empire**

Love versus Empire

(the beginning of the end of Empire)

Love versus Empire - the beginning of the end of Empire

While Napoleon wasn't aiming directly for genocide with his campaign

Napoleon and his staff at Borodino by Vasily Vereshchagin

While Napoleon wasn't aiming directly for genocide with his campaign, such as the modern campaigns now do with the biofuels hoax, Napoleon was nevertheless a high-minded imperialist who didn't care the least how many people would be killed for the expansion of the system of empire that he served. Neither did Adolf Hitler care any more in later times, and Stalin before him, when they blew the trumpet for genocide. And why should they have cared about human living and happiness, when the masters of empire had trumpeted loud and clear from the late 1700s on that the world is woefully overpopulated, and needs to be 'cleansed' of its 'human pest' as they had repeatedly announced with a smile and still do.

Mass genocide isn't deemed a crime in the circle of the high-minded who use people up like dumb animals in pursuing their political objectives. War, thereby has never been a crime in their eyes, but an ideal, a way of doing business. The fundamental legality of war had been long established in the highest halls of empires with numerous names. In modern time this fundamental legality has been expanded to now include the legality to inflict genocide at will, and depopulation as desired. And in very recent times the further legality has been added, to unleash a first-strike nuclear war; unprovoked; at will; in the service of imperial objectives.

The genocidal objective that started with the doctrine of the Empire Sun, has become a monster

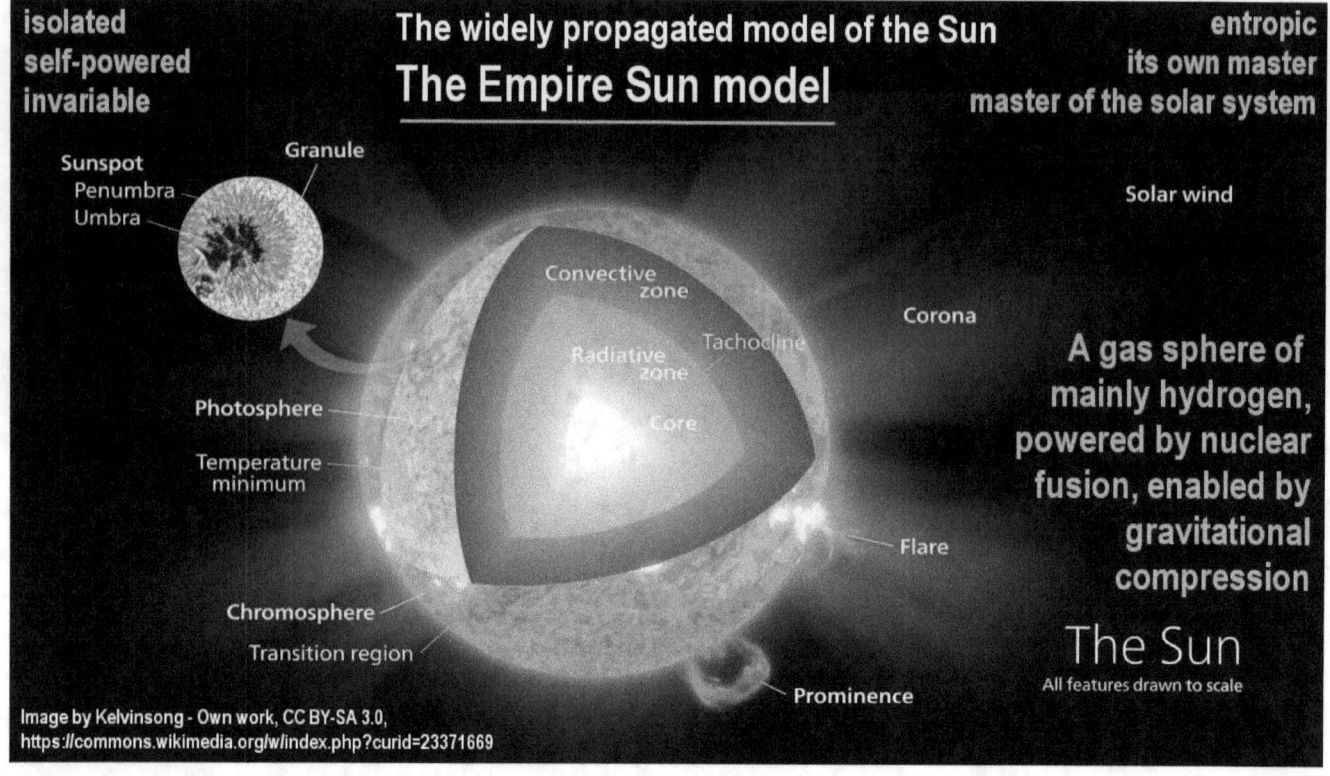

In other words, the genocidal objective that started with the doctrine of the Empire Sun, has become a monster in the modern world. Nevertheless, the overturning of this basic doctrine may serve as a good start towards the freedom of humanity and its having a future.

Napoleon was a saint in comparison with modern society

Napoleon was a saint in comparison with modern society, in that he only killed half a million men in half a year, while the West in modern time kills a hundred million people by starvation every single year with the biofuels hoax without even raising an eyebrow.

Napoleon didn't intend the tragedy that he caused to happen

Napoleon and his staff at Borodino by Vasily Vereshchagin

Napoleon didn't intend the tragedy that he caused to happen that his war became, while we in our time do fully intend the vastly greater genocidal tragedies, by policy, loosely termed depopulation. Napoleon wanted to build a land bridge to India. That's what he had aimed for. He just failed to realize that nothing is ever accomplished with war and the destruction of society. He failed to recognize the real gold that civilization depends on, which is the creative power of society's humanity.

In contrast with Napoleon, who wanted to develop a land bridge to India

In contrast with Napoleon, who wanted to develop a land bridge to India, the western empire in modern time wages war in evermore places to prevent economic development, worldwide, from becoming possible. The modern goal of empire, which the entire imperial West is locked into, is: No progress, No development, No industrial revolutions. In the name of empire, the West hails fascism, depopulation, food burning, terror, and poverty instead.

In comparison with our ideals in modern time, the mad dog Napoleon was a saint indeed. And to top off the account of modern madness, society actively closes its eyes to the onrushing near Ice Age for which the fringe effects are already mounting up evermore.

Napoleon's troops met their ice age that almost none survived

Napoleon's troops met their ice age that almost none survived, just as modern society won't survive unless the logistics are in place that enable humanity to live and prosper in the coming Ice Age World.

We, in our time, are in need to build a World Bridge with Africa at its center

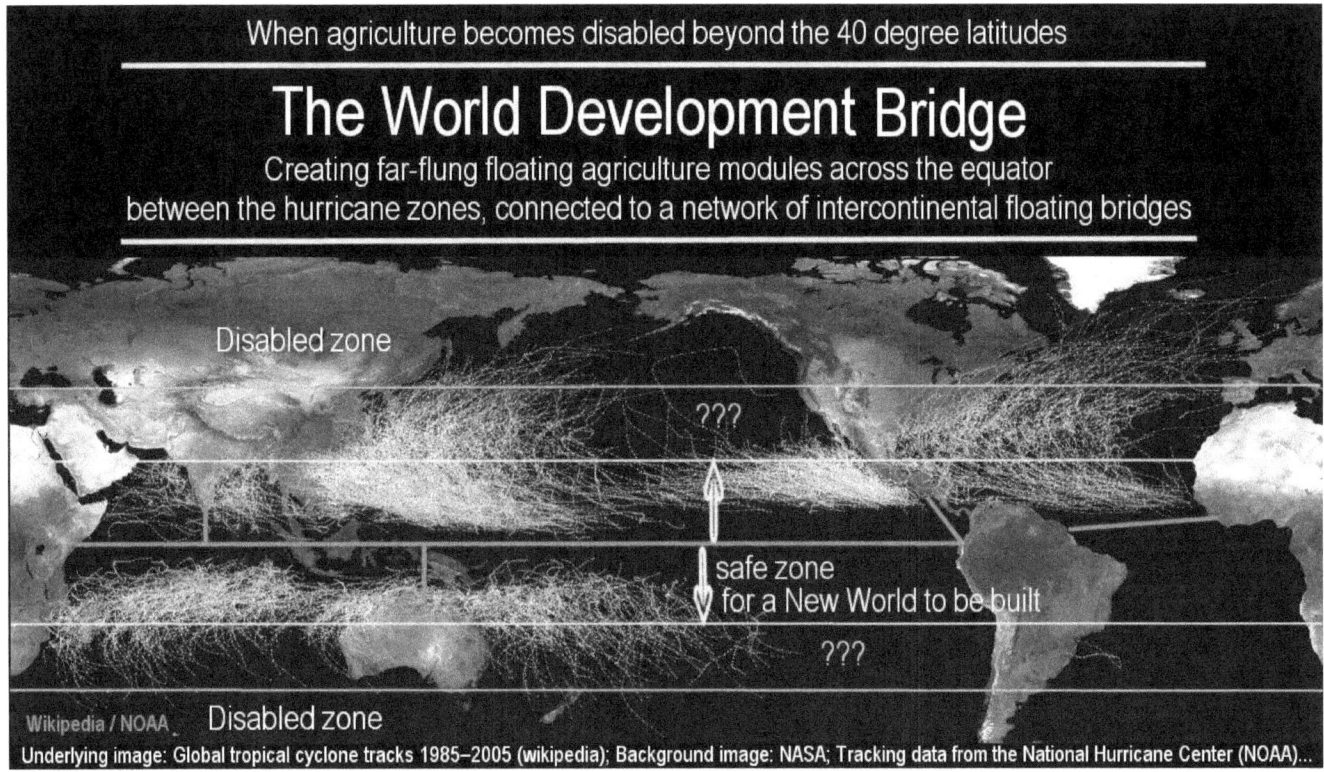

Napoleon wanted to build a land bridge to India. We, in our time, are in need to build a World Bridge with Africa at its center, and re-develop Africa into the logistical hub of the world for its equatorial location, together with equatorial America. The habitable world is poised to be shrinking to the narrow band of the tropics. The process has already begun. The phase shift will happen in the 2050s or or earlier. Will we be ready for it?

No one can answer this question. One can only answer with certainty that nothing will be built if we don't get started. Our present commitment on this line is to close our eyes, remain asleep, and do nothing.

Napoleon's soldiers would cry out to us from their icy grave if they could

Napoleon's soldiers would cry out to us from their icy grave if they could. They would be crying, 'you fools, open your eyes, your sleeping will kill you. Awake and live as human beings while you still have the chance.'

- **Divine Love, reflected in universal love for one another**

Divine Love

reflected in universal love for one another,
always has met and always will meet every human need

This chance rests on a principle, a great principle. The principle is:

Divine Love, reflected in universal love for one another, always has met and always will meet every human need.

Unfortunately there is little of this love reflected anymore in many of the once great nations. The stage is almost empty. Its deemed value is near zero.

In 1812 the soldiers did not wield their swords out of love for humanity, but to kill

The Grande Armée crossing the Niemen River into Russia, June 1812

In 1812 the soldiers did not wield their swords out of love for humanity, but to kill. By this intention civilization is doomed, as the soldiers all had experienced. The victory was on the side of those who had chosen to live. Their victory wasn't won until the next year.

It may have been the frivolous waste of the 680,000 of Europe's most able men

It may have been the frivolous waste of the 680,000 of Europe's most able men that were lost in Russia, and 300,000 of its finest horses, all in the pursuit of Napoleon's wild dreams of expanding the empire all the way to India, that a coalition began to form against Napoleon and the system of Empire that he personified. The coalition comprised Russia, Austria, Prussia, Sweden, Great Britain, Spain, Portugal, and some smaller German states.

Miraculously, Napoleon managed to mobilize a new army against the Alliance

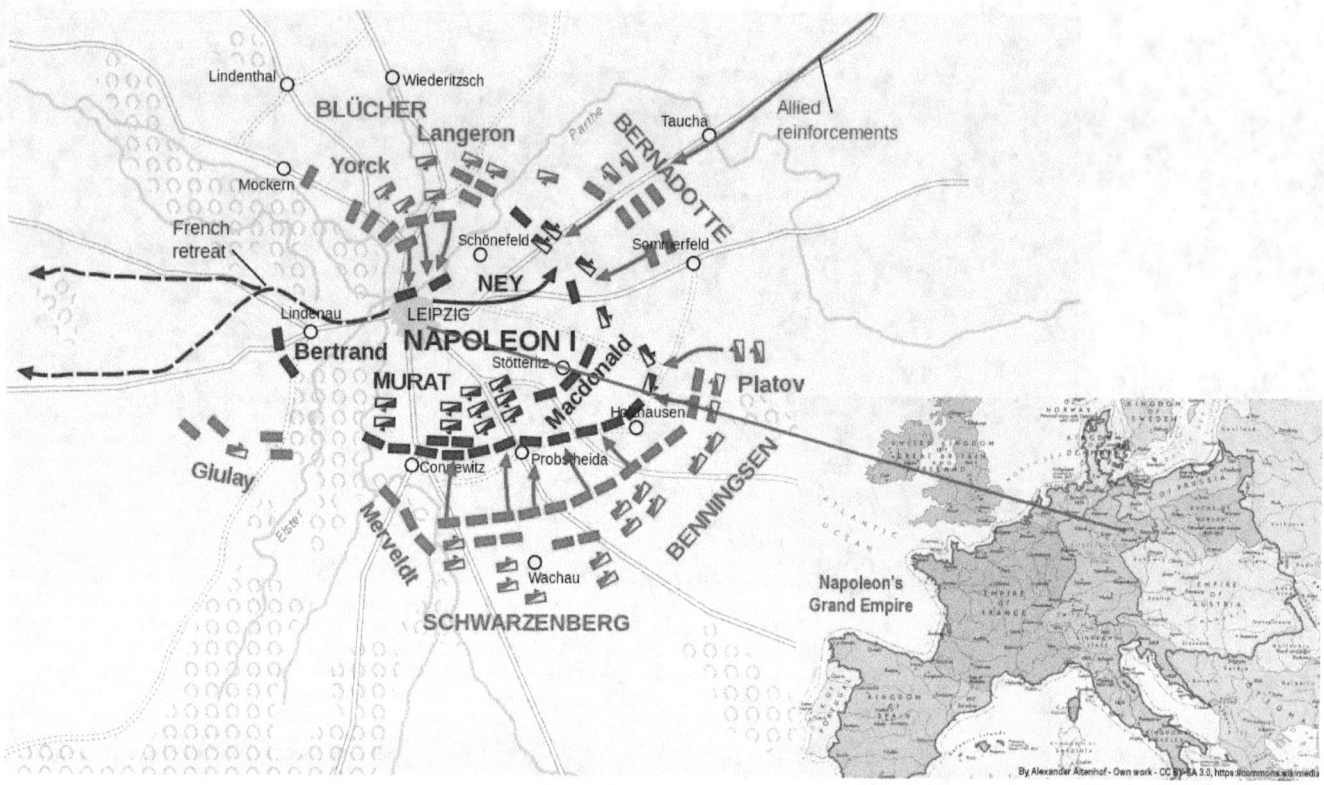

Miraculously, Napoleon managed to mobilize a new army against the Alliance, of about the size of the one he had wasted in Russia. However, the waste of the human potential that by then was history could not be undone. The fascism of the empire had changed the landscape against it. The flow of history was irreversibly turning against what Empire stood for.

The decisive battle that marked the end of the vast French Empire was centered on the City of Leipzig.

The French-Empire forces were already loosing before the big battle happened

The French-Empire forces were already loosing before the big battle happened. The French began to be routed in Spain and Portugal with the aid of the British on the 9th of October in 1813.

The Battle of Leipzig, in Germany, started a week later

The Battle of Leipzig, in Germany, started a week later. It became known as the Battle of the Nations.

The battle was brought into the city

The battle was brought into the city.

The position at Leipzig afforded major advantages to Napoleon's army

The position at Leipzig afforded major advantages to Napoleon's army and his battle strategy.

Several rivers converged and split the surrounding terrain into separate sectors

Several rivers converged at and near the city. The rivers split the surrounding terrain into separate sectors.

By holding the City of Leipzig and its nearby bridges, Napoleon could shift troops from one sector to another more rapidly than could the Allies against him.

The Allies had difficulty moving large numbers of troops into Napoleon's sectors

The Allies had difficulty moving large numbers of troops into Napoleon's sectors across the rivers.

But the momentum was no longer on Napoleon's side

But the momentum was no longer on Napoleon's side. The die was cast against him - against Empire.

After 4 days of war, on October 19, Napoleon ordered the retreat

After 4 days of war, on October 19, Napoleon ordered the retreat. The Allies eventually pursued him across the river Rhine.

History records little of the details, of the struggles, hopes, sacrifices

History records little of the details, of the struggles, hopes, sacrifices. It records only the number of dead.

History also records that the vast French Empire existed no more

History also records that the vast French Empire existed no more after the smoke blew away with the wind.

The Empire lost this day all of its territories east of the Rhine

The Empire lost this day all of its territories east of the Rhine. Napoleon, himself, was forced to abdicate. He was exiled in 1814.

The liberated people who had won their freedom from Empire

The liberated people who had won their freedom from Empire, constructed a giant monument to commemorate the achievement. To judge by the size of the monument they regarded their liberty a momentous achievement.

Construction began 84 years later in 1898. The work was completed in 15 years.

As early as 1814 proposals were made to build a monument to commemorate the victory over Empire. Construction began 84 years later in 1898. The work was completed in 15 years. It stands 299 feet tall. That's more than half as tall as the Great Giza Pyramid in Egypt.

The great monument of freedom from Empire is constructed in granite-faced concrete

The great monument of freedom from Empire is constructed in granite-faced concrete. Both granite and concrete have the quality of endurance.

The power that dissolves Empire is the humanity of the human being.

The monument's spacious interior, unlike the chambers of the pyramids, features gigantic sculptures of the humanity of the nations who fought for their freedom and won. The power that dissolves Empire is the humanity of the human being.

The monument is also known for the wide view that it offers to visitors

The monument is also known for the wide view that it offers to visitors. Perhaps this is symbolic, because when society stands above Empire, the view of its future unfolds with a near infinite horizon. In Science, this is the horizon that opens up beyond the system of Empire and the model of the Empire Sun.

The very future of humanity depends on us gaining the freedom of this wider horizon in our time.

The year 1812 marks the beginning of the end of the system of Empire

The year 1812 marks the beginning of the end of the system of Empire as a political structure. The year thereafter, 1813, marks a decisive step forward in Europe, although the war against the system of empire had not yet been won at this time; nor has it been won to the present day. The system of empire still rules the world, politically, financially, militarily, and especially scientifically with devastating effects.

One day, however, and hopefully soon, the deadly poster of the Empire Sun

One day, however, and hopefully soon, the deadly poster of the Empire Sun may find its final resting place in future museum displays of a dark stage in history.

Till then, the poster of the Empire Sun still rules and humanity remains in crisis

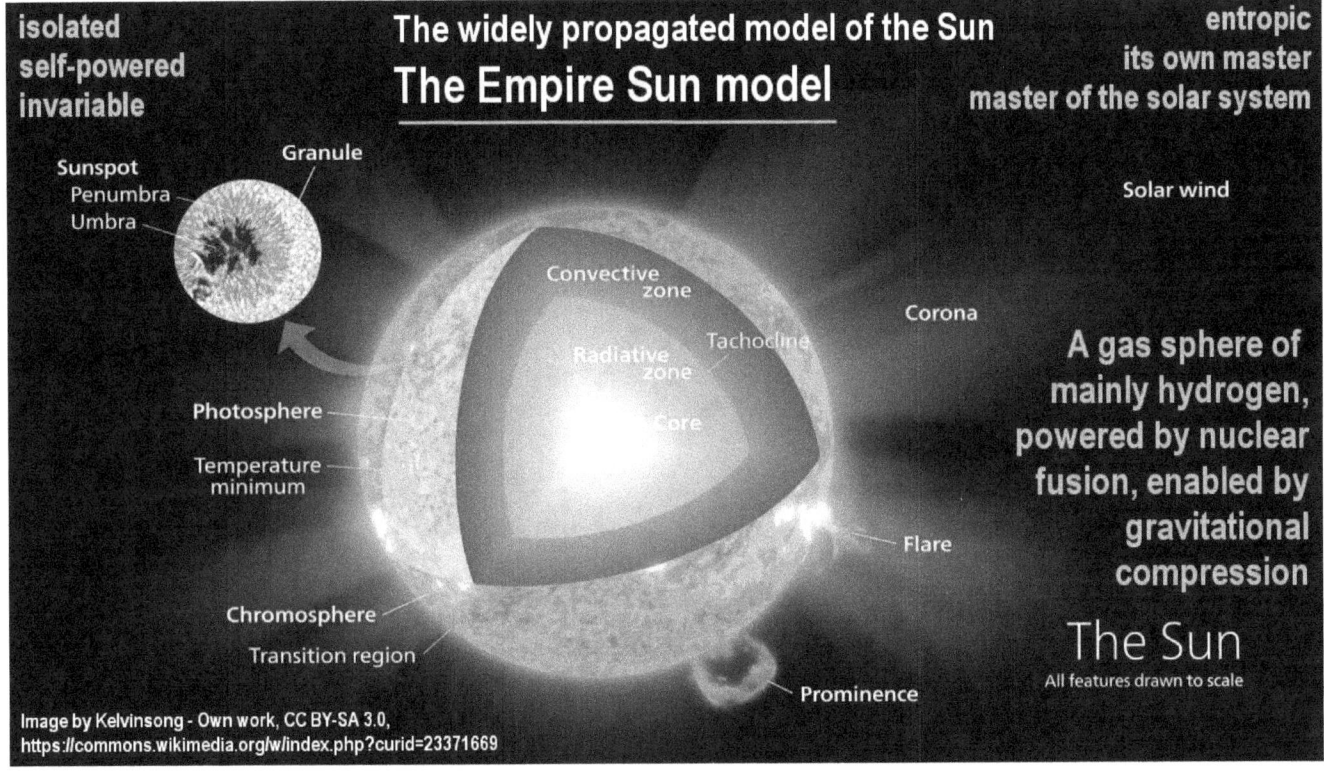

Till then, the poster of the Empire Sun still rules and humanity remains in an existential crisis as it cannot see what remains hidden from it by its smallness in thinking. If this doesn't change soon, the rule of the false poster will destroy nearly the whole of humanity with the Ice Age phase shift that it then meets unprepared for the consequences.

However, times are changing

However, times are changing. The great principle, Universal Love, the principle that stands as the opposite of empire, remains standing, waiting to be picked off the ground and to be held high. And this too, is happening to some degree already.

Some strong movements have begun in this direction, led by China.

If the G20 logo symbolizes solar plasma-flow physics, it would indicate

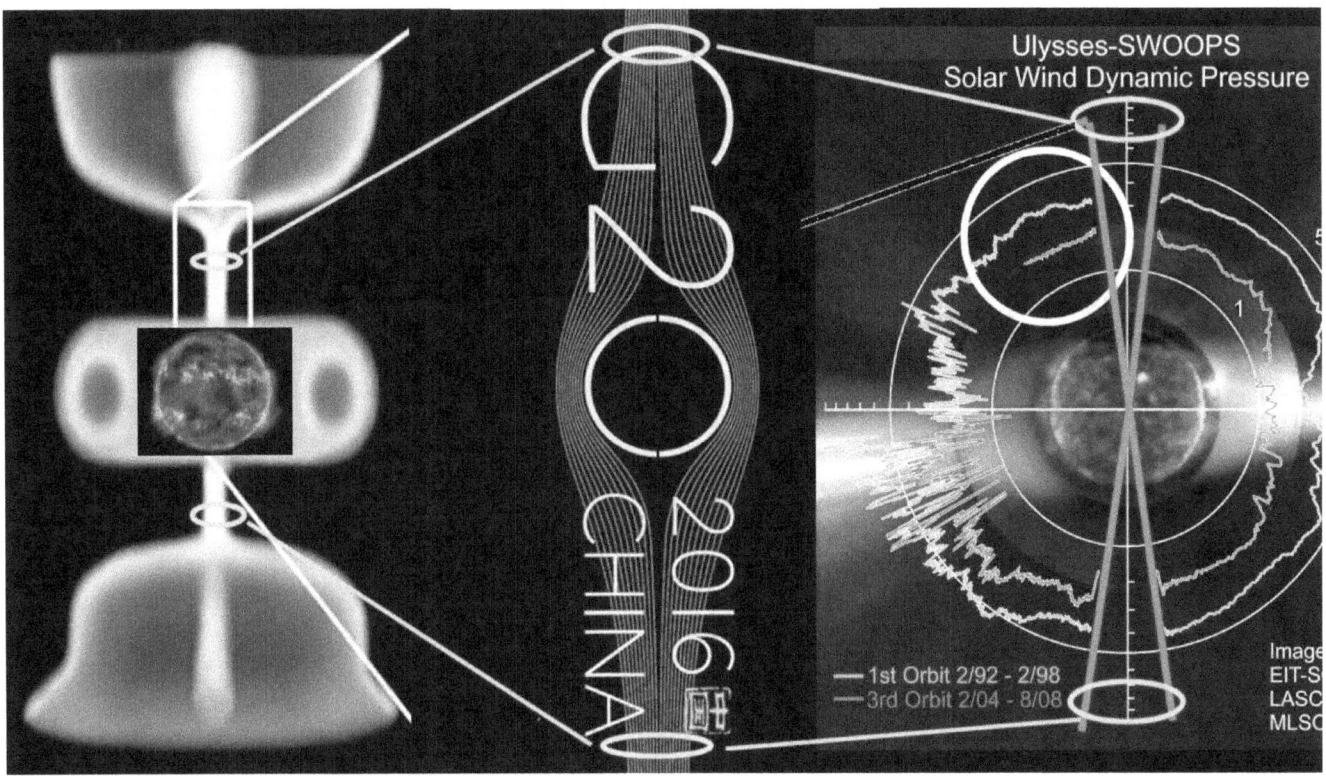

If the G20 logo symbolizes solar plasma-flow physics, it would indicate that China is well aware of the great challenge before it and before us all.

> **Ice Age = Opportunities**

| Ice Age = Opportunities |

Ice Age = Opportunities

Whether humanity succeeds on this front remains yet to be seen

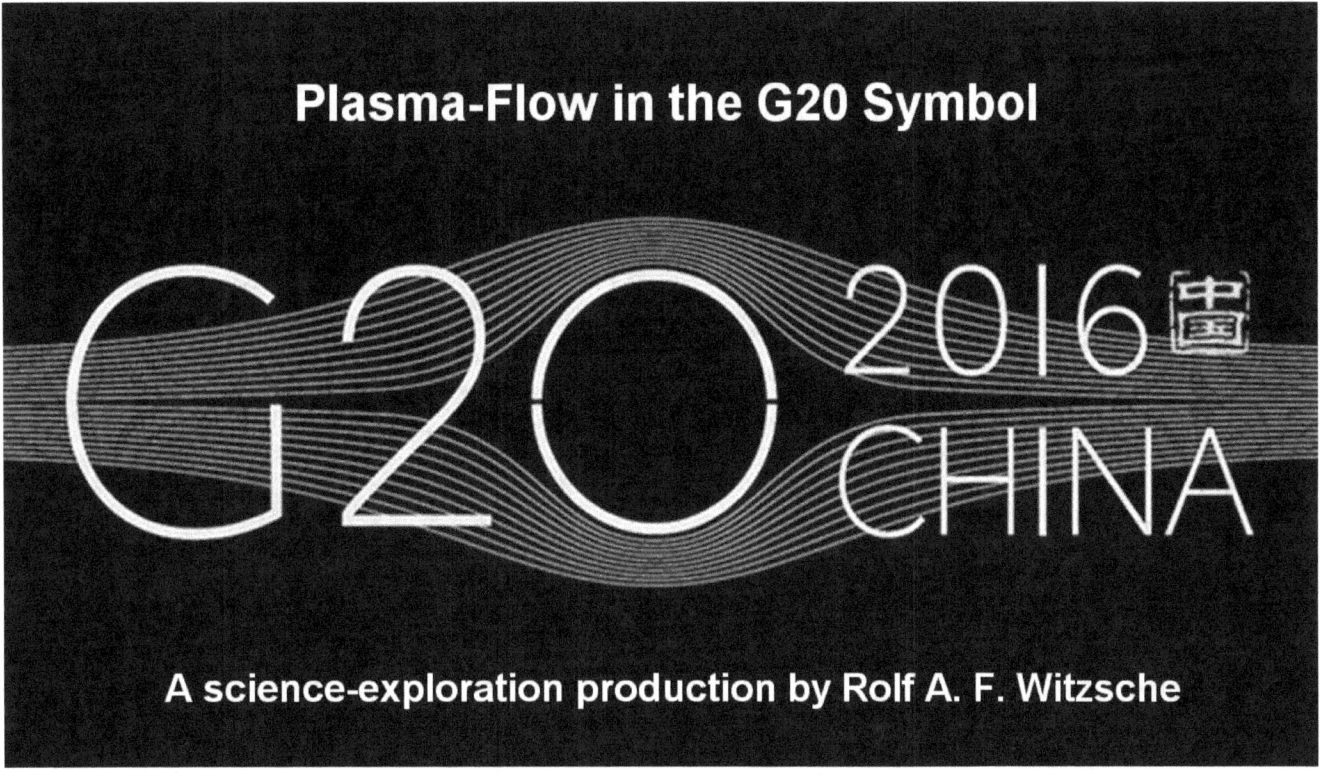

Whether humanity succeeds on this front, working together across the world as one people, in love with our universal humanity, and builds itself the needed infrastructures in preparation for the near Ice Age, in order for us all to live and prosper therein, and not perish, remains yet to be seen.

If humanity fails to break itself away from the still widely accepted doom

If humanity fails to break itself away from the still widely accepted doom that it is exposed to under the White Spider of Empire and its false Sun, then the coming Ice Age will find humanity unprepared for it, which will thereby be its final demise that it won't recover from for a very long time.

The time has come for humanity to take the trash out of the house

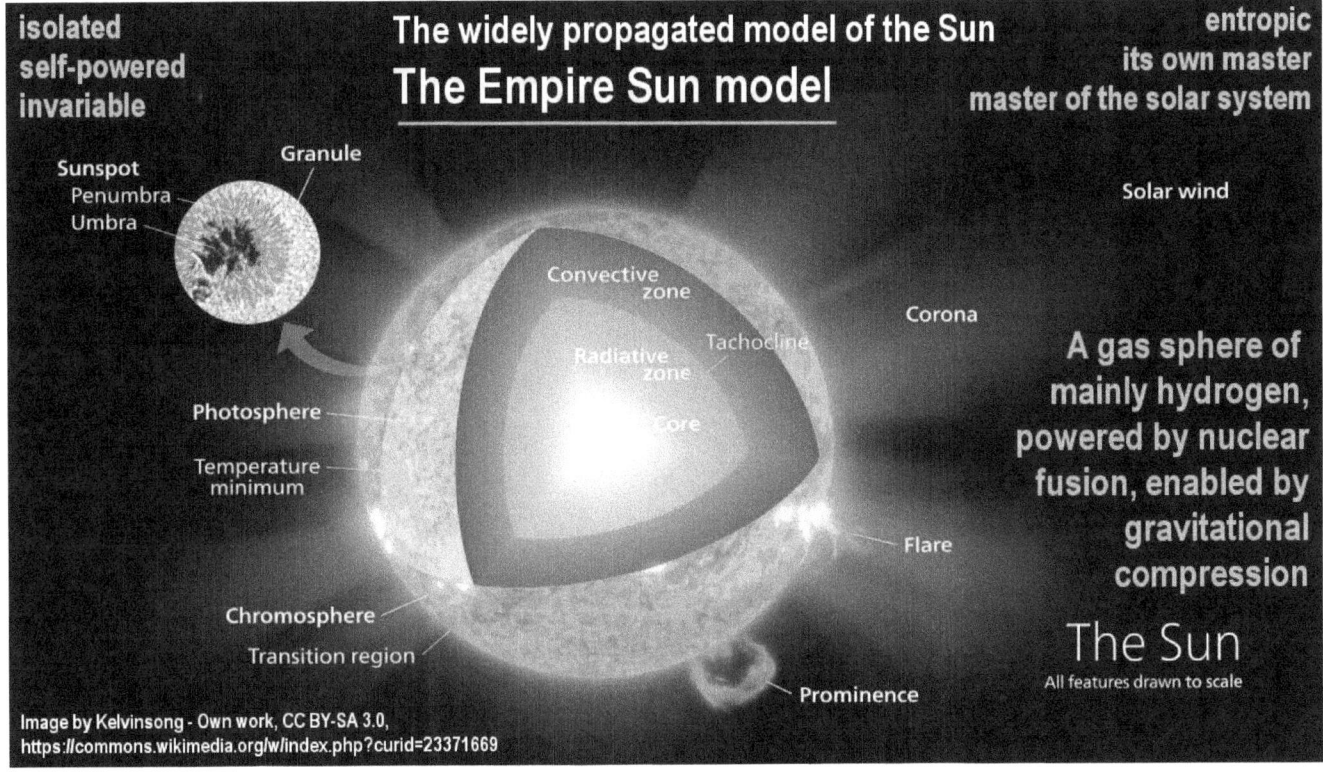

This doom is is prepared for under the banner of the Empire Sun.

The time has come for humanity to take the trash out of the house so that it can live again, and develop, and have a future in a clean house. Proposals were made as early as 1814 to build a monument to commemorate the victory over Empire, which the Leipzig monument became.

Its humanity still stands

Its humanity still stands.

Its aspiration for freedom still stands

Its aspiration for freedom still stands.

Its vision still stands and continues to enrich the landscape

Its vision still stands and continues to enrich the landscape.

The victory that it commemorates was won in 4 decisive days in 1813

The victory that it commemorates was won in 4 decisive days in 1813. The building of the monument was the work of 15 years, paid for mostly by donations, so significant was the victory regarded in society. It was inaugurated in the presence of 100,000, to celebrate the 100th anniversary of the event in 1913. It is built on the spot where Napoleon ordered the retreat of his army. It still stands today as a commemorative symbol, after more than a century has passed.

The Statue of Liberty in New York harbour stands on the same eternal foundation

The Statue of Liberty, completed in 1886, representing the Declaration of Independence, America's fundamental constitution that has stood by then, for 110 years.

The Statue of Liberty in New York harbour stands on the same eternal foundation, of the quest for liberty from Empire. It was dedicated 27 years before the Leipzig monument was dedicated, and stands roughly half as tall, but not in significance.

The copper statue was donated as a gift from the people of France

The copper statue was donated as a gift from the people of France to the people of the United States for the same type of celebration. The statue was designed by French sculptor Frédéric Auguste Bartholdi and built by Gustave Eiffel. The project was traced back to a 1865 conversation between Édouard René de Laboulaye, a staunch abolitionist and Frédéric Bartholdi, a sculptor. In conversation, Laboulaye, a supporter of the Union in the American Civil War, is quoted to have said: "If a monument should rise in the United States, as a memorial to their independence, I should think it only natural if it were built by united effort—a common work of both our nations."

➢ Where the heart smiles, there is liberty

Where the heart smiles
there is liberty

Where the heart smiles, there is liberty.

The pedestal on which the statue stands was paid for by 120,000 individual contributions

Statue of Liberty unveiled, by Edward Moran - 1886 - wikipedia

Liberty is the expression of sovereignty established in the heart.

The pedestal on which the statue stands was paid for by donations in the form of 120,000 individual contributions. This echoed its significance.

It is said that with the abolition of slavery and the Union's victory in the Civil War in 1865, Laboulaye's hopes for freedom and democracy became reality in the United States, so that in honouring the achievement, Laboulaye proposed that a gift be built for the United States on behalf of France, perhaps secretly hoping that by calling attention to the achievement in the United States, the French people would be inspired to reach for their own freedom from a repressive monarchy, that of Napoleon III. The French monarchy was abolished in 1870 after Napoleon III was captured in one of his wars.

Since then the gaggle of the Empire institutions began to shrink. More and more monarchies were abolished. Fascist intuitions rose and continued the system of empire under new colors, and were abolished too in turn. This does not mean that the end of Empire was near at the time. Empire would continue, and so would its wars, even bigger wars, but the beginning of the end had begun.

This beginning of the end of Empire may have been recognized already in 1813

This beginning of the end of Empire may have been recognized already in 1813 when a decisive step was made towards freedom and love. The opposite of Empire is freedom and love. Wars will cease on this path. Wars are not instigated by love, neither are stealing, terror, and genocide. A sea change against these had begun by the once again rising tide of love. The healing is carried by universal love

The Statue of Liberty prefigures the inevitable recognition of the inherent dignity

The Statue of Liberty prefigures the inevitable recognition of the inherent dignity and unity of all mankind that remains standing tall for everyone to see and to aspire to. And so, the quest continues for freedom in thinking and freedom in science.

➢ **We are not impotent - this is true now**

We are not impotent

(this is true now)

We are not impotent - this is true now

While the word, 'Ice Age,' is no longer spoken, as it falls outside the framework

While the word, 'Ice Age,' is no longer spoken, as it falls outside the framework of the politically correct, the truth still beckons; just take me in, uplift yourself, protect yourself.

We need a 'university' type approach here, in order to bring the truth back into view, which false posters have hid for far too long. The very idea of protecting and developing humanity and its food resource by meeting the greatest challenge in human history that is now upon us, is effectively slandered so that it is almost deemed a crime to entertain the concepts of protection, industry, and development.

This debilitation can end. The landscape can be changed. While the looming Ice Age cannot be avoided, the tragedy that much of humanity is rushing into by not preparing itself for it, can be avoided. The door is still open, though its timeframe is shrinking.

The Ice Age Challenge before us should be the banner headline in today's world

The Ice Age Challenge before us should be the banner headline in today's world. Instead, the subject is shunned even in Russia and China, although the leaders there appear have begun to orient themselves towards creating a future for their nation and for humanity.

The masters of empire, in contrast, still call it a crime

The masters of empire, in contrast, still call it a crime - a crime against empire - for society to speak about protection, industry, and development, and its aiming to live abundantly.

Under the shadow of their golden pumpkin that society adores, of which there are now many types, the heart of civilization has become so isolated and desecrated, that Empire has become a way of thinking, of small-minded thinking, the kind of thinking that Napoleon exemplified, in comparison with what society has the potential to be, and may yet be.

Thus the time for decision is at hand: a time for humanity to become human in the fullest sense. The time has come for society to rediscover its humanity as human beings.

The critical success in this arena depends on whether humanity raises itself up

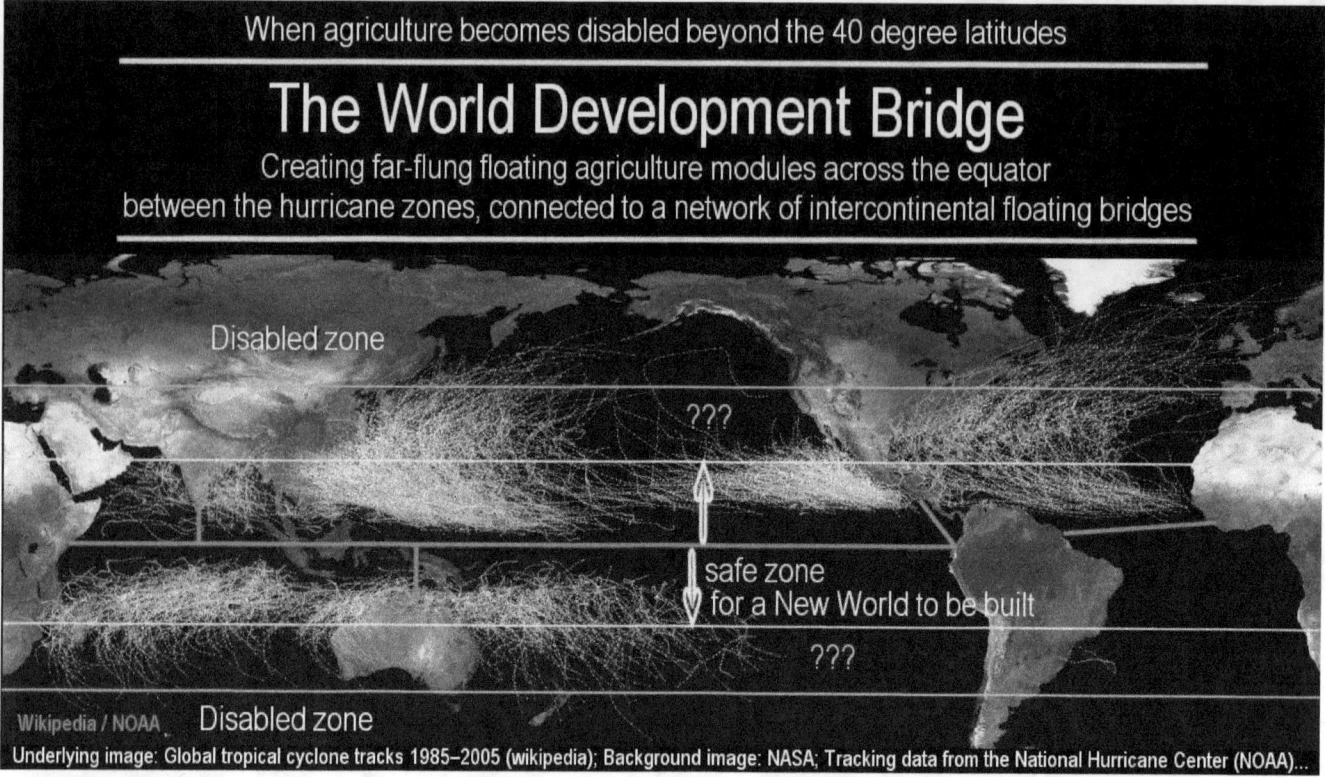

The critical success in this arena depends on whether humanity raises itself up sufficiently to reach for the one great principle that everything depends on in civilization, which is the Principle of Universal Love, and raises this principle up high.

I think this can be done. With this we can succeed.

To succeed on this front is well within the range of what can be accomplished by a society of human beings.

➤ **Arise and Shine our time has come - humanity=infinity**

Arise and Shine

our time has come
(humanity=infinity)

Arise and Shine our time has come - humanity=infinity

Like seeds are we, wind-blown

Like seeds are we, wind-blown
Carriers of a secret still unknown
Poems in the words of nature
Sentinels of an Intelligence yet unseen
Prophets of the enduring
Apostles in an endless landscape

Harvest is seedtime, thoughts ripening

Harvest is seedtime, thoughts ripening
Carried as by a great wind
Carriers of secrets to unfold
Thoughts winged with Purpose
A force waiting, silent
Patiently waiting for the moment

Thoughts do awaken

Thoughts do awaken
Roused by the moist warmth in spring
Cascades of colors, colors of life
Bright yellows, bursts of silver-white
Thoughts becoming creations
Monuments of genius, builders of worlds

Who owns the seeds? Do we?

Who owns the seeds? Do we?
Who can fathom their wonder?
Life flows from then in great rivers
Rivers trailing into oceans
In them we are alone, each one is alone
Each thought is sovereign, beauty is its song

Thoughts are seeds, becoming ideas

Thoughts are seeds, becoming ideas
Alive in discovering
Alive in listening
Alive in being touched by love
Alive in loving
Alive...

Like seeds, thoughts fall to the ground

Like seeds, thoughts fall to the ground
Potentials are lost
Hard grounds kill the precious
But we are Man
Hard ground becomes tilled, watered
The precious is nurtured in loving

Love for one-another, the human spring

Love for one-another, the human spring
Mankind is afloat in a sea that is Love
Seeds germinate, become plants
Roots break the ground
Love lifts the barriers, patiently
Silently waiting, reaching for the sky

Thoughts are the Universe unfolding

Thoughts are the Universe unfolding
Landscapes of brilliance, ideas of power
Substance for enriching one-another
Substance of the forever maturing
Thoughts bearing new seeds within
Seeds for splendors beyond dreams

Each harvest is seedtime

Each harvest is seedtime
A seed becomes a plant bearing new seeds
A thought unfolding, bears up civilization
A spark in the heart, bears the 'fire' of life
New worlds are created in the 'fire' of passion
We are the bearers of a 'fire' that is light

Builders of worlds are we

Builders of worlds are we
New Worlds, which have never been
Precious with riches grander than our own
Nature is Love reflected in loving
Love paints with the colors of its endless spring
Love paints us all - but who owns the seed?

Who owns the cradle for the seed?

Who owns the cradle for the seed
Name it Intelligence, name it the Universe
Thoughts are seeds from an infinite fountain
Monuments of grandeur of good
Fields of flowers dancing in the sunshine
All nature whispers this to us

The melody of nature - what a song!

The melody of nature - what a song
Whispers of a splendour grander than the heavens
Like seeds are we - we whisper too
Seeds bearing gifts for the world
Gifts wrapped up in sunshine
Gems are we - unfolding a majestic song!

listen to the song

Listen to the song
Listen to the heart
Listen to the silence where strands of love unfold

Listen to the symphony of our humanity

Listen to the symphony of our humanity
In this symphony we are One
One with the Universe itself.

From the poem, Harvest is Seedtime, by Rolf A. F. Witzsche
Celebrating the End of the Beginning - The Future is what we make it to be

----

This exploration video is presented in honour of
the May 14-15, 2017 world summit in Beijing, China:

➢More from the author:

14 Libraries of books and video productions

Novels on Universal Love, the greatest principle in civilization - **14 major novels**

    Flight Without Limits (science fiction)

    Brighter than the Sun (nuclear war avoidance?)

    A series of twelve novels: **The Lodging for the Rose** exploring the Principle of Universal Love

    Book 1 - **Discovering Love**

    Book 2 - **The Ice Age Challenge**

    Book 3 - **Roses at Dawn in an Ice Age World**

    Book 4 - **Winning Without Victory**

    Book 5 - **Seascapes and Sand**

    Book 6 - **The Flat Earth Society**

    Book 7 - **Glass Barriers**

    Book 8 - **Coffee Sex and Biscuits**

    Book 9 - **Endless Horizons**

    Book 10 - **Angels of Sex in Queensland**

    Book 11 - **Sword of Aquarius**

    Book 12 - **Lu Mountain**

The Sex and Sacrament Project - exploration stories from my novels - **11 books**

    The Son of God

    Impotence and Power

    Self-Love and the Golden Hijab

    Erica's Flower Garden

    Helen a Healer

    Brilliance of a Night

    Gem of the Universe

    The Sound of a Bird Woke Me

    Between Ice and Spirit

    Anton of Grace

    Goodness of Living

The Kaleidoscope Project - mixed media of stories from my novels
- videos, PDF, audio

Discovering Infinity - developing history - 13 major research books:
A Research Book Series focused on scientific and spiritual development

    Volume ii (Introduction) **Roots in Universal History** (Focus on Reality)

    Volume 1A **The Disintegration of the World's Financial System** (Focus on Truth)

    Volume 1B **Crimes Against Humanity** (Life Denied)

    Volume 2A **Science and Christian Healing** (History as Truth)

    Volume 2B **The Lord of the Rings' Metaphors**

    Volume 3A **Universal Divine Science: Spiritual Pedagogical** (Structure for Discovery and Scientific Development - The Scientific Process to Know the Truth)

    Volume 3B **Science and Health with Key to the Scriptures in Divine Science**

    Volume 3C **Bible Lessons in Divine Science - 1898**

    Volume 3D **Living in the Sublime**

    Volume 4 **Light Piercing the Heart of Darkness** (The Demands of Truth and Justice)

Volume 5 **Scientific Government and Self-Government** (Platform for Freedom)

Volume 6A **The Infinite Nature of Man** (The Fourth Dimension of Spirit)

Volume 6B **Leadership** (The Spiritual Dimension of Leadership)

Cool Science of Kids - Illustrated Science - **interactive, videos, and 20 books**

War, Economics, and Nuclear War - scientific exploration - **10 videos**

Civilization - series focused on humanity - **10 videos**

Global Warming Doctrine - science videos - **12 videos**

Freshwater and Energy - science videos - **7 videos**

Christian Science explorations - **16 videos**

Books by Mary Baker Eddy - Christian Science - **16 on-line books**

Books by Rolf Witzsche on Christian Science - **9 Books**

The Giant PDF Library all transcripts of videos in PDF form

For links, please see: http://www.ice-age-ahead-iaa.ca

The projects are designed to draw the riches of our humanity into the foreground **towards a New Renaissance**, in order that their light may out-shine the systems of empire that are erroneously accepted, including the follies of war, terror, looting, economic destruction, science-perversion, and policies for depopulation.
Rolf A.F. Witzsche

www.ingramcontent.com/pod-product-compliance
Lightning Source LLC
Chambersburg PA
CBHW062350220526
45472CB00008B/1759

9 781719 244374